ONLINE SERVICE BUSINESS

CONSUMER BEHAVIOUR AND COMPETITIVENESS

ONLINE SERVICE BUSINESS
CONSUMER BEHAVIOUR AND COMPETITIVENESS

BY

ROBERT LYNN
SHERRY DALSEY

KNOWLEDGE BAKERS

Cataloging in Publication Data--DK

Courtesy: D.K. Agencies (P) Ltd. <docinfo@dkagencies.com>

Lynn, Robert (Writer on online service business), author.

Online service business : consumer behaviour and competitiveness / by Robert Lynn, Sherry Dalsey.

pages cm

Includes bibliographical references.

ISBN 9789349964259

1. Internet banking. 2. Consumer behavior. I. Dalsey, Sherry, author. II. Title.

LCC HG1708.7.L96 2023 | DDC 332.102854678 23

ISBN: 978-93-49964-25-9

Published by : **Knowledge Bakers**
Pune, India
Email: knowledgebakers2019@gmail.com

Preface

Consumer intentions to purchase services, particularly financial services, traditionally depend on intangible aspects of quality associated with face-to-face interactions between buyers and sellers. However, electronic commerce in general, and Internet retail banking in particular, ostensibly removes most human contact from service provision. This has profound implications for services marketing, and research methodology in particular. These observations concerning Internet based services, such as banking, give rise to a fundamental question; what characteristics of an Internet banking service influence consumer purchase intentions?

Following an extensive literature review from a number of apparently disparate yet intrinsically interconnected areas of study, it is hypothesised that consumer purchase intentions for Internet banking is dependant upon information search and experience qualities, functionality of accounts, branding aspects, service quality aspects, price and security.

In these novel circumstances, this thesis takes a first step toward understanding these implications for marketing research, by drawing attention to the methodological opportunities offered by Internet banking services. This study moves away from the traditional attitudinal research that has been used in previous research, and develops mystery-shopping research methodology as an innovative alternative approach to understanding consumer purchases in Internet banking services. Accordingly a series of Internet mystery shopper scorecards are developed as the research instruments.

Problems with data render full hypothesis testing infeasible at this stage. Nevertheless, an exploratory analysis of fifteen Internet banking services operating in the UK as of January 1, 2000 illustrate the potential methods of analysis using the proposed research methodology. Key components for Internet services marketing success tentatively identified by this research are price, time elapsed since launch, parent brand association, service quality and security.

Robert Lynn
Sherry Dalsey

Acknowledgements

The most pleasant task in this research work is expressing my appreciation to the many individuals who helped me complete this work. When I began, I had no idea that the scope of the job that would require me to request, plead, cajole, and charm a number of people into helping me. It is difficult to list all those individuals and therefore I start out by expressing my general indebtedness and enormous gratitude to everyone whose invaluable assistance has contributed to the fruitution of this thesis.

And now on to the specifics ...

I am indebted to my Director of Studies and first supervisor, Professor Paul Foley, Director of IECRC, for his patience, encouragement and guidance throughout the preparation of this thesis. I am equally grateful to my other supervisors Professor Trevor Buck, for his deep sense commitment and constructive criticism and to Brahim Herbane for his continued support.

Last but not least, I wish to thank De Montfort University for providing financial support for my studies.

Robert Lynn

CONTENTS

LIST OF FIGURES

LIST OF TABLES

Abbreviations

1.	AARNET	Australian Academic Research Network
2.	ACD	Automated Call Distribution
3.	ADSL	Asymmetrical Digital Subscriber Lines
4.	ARPA	Advanced Research Projects Agency
5.	ARPAnet	Advanced Research Projects Agency Network
6.	ATM	Automated Teller Machine
7.	AUP	Appropriate Use Policy
8.	BARRNET	Bay Area Regional Research Network
9.	BBA	British Bankers Association
10.	BITNET	Because It's Time NETwork
11.	BOS	Bank of Scotland
12.	BPR	Business Process Engineering
13.	CAD	Computer Aided Design
14.	CEARN	CEuropean Academic and Research Network
15.	CEO	Chief Executive Officer
16.	CD-R	Compact Disk Recordable
17.	CERFnet	California Education and Research Network
18.	CERN	Centre for European Academic Research Network
19.	CERT	Computer Emergency Response Team
20.	CFO	Chief Finance Officer
21.	CIX	Commercial Internet eXchange
22.	CME	Computer Mediated Environment
23.	CNRI	Co-operation for the National Research Initiative
24.	CREN	Co-operation for Research and Education Networking
25.	CSNET	Computer Science Network
26.	DNS	Domain Name Server

27.	DoD	Department of Defence
28.	EAI	Enterprise Application Integration
29.	E(e)-commerce	Electronic commerce
30.	EDI	Electronic Data Interchange
31.	EFF	Electronic Frontier Foundation
32.	E(e)-mail	Electronic Mail
33.	ERP	Enterprise Resource Planning
34.	EUnet	European UNIX Network
35	FTP	File Transfer Protocol
36	GIS	Global Information Solutions
37	GOSIP	US Government OSI Profile
38	GUI	Global User Interface
39	HSBC	Hong Kong and Shanghai Banking Corporation
40	HTML	Hyper Text Markup Language
41	IAB	Internet Activities Board (later Internet Architecture Board)
42	ICCB	Internet Configuration control Board
43	IE	Internet Explorer
44	IF	Intelligent Finance
45	IP	Internet Protocol
46	IRC	Internet Relay Chat
47	ISA	Individual Savings Accounts
48	ISDN	Integrated Services Digital Network
49	ISOC	Internet Society
50	ISP	Internet Service Provider
51	JANET	Joint Academic Network
52	JIPS	JANET IF Service
53	LAN	Local Area Network
54	Mac	Apple Macintosh
55	MEG	Moderated Email Groups
56	MRS	Market Research Society
57	MUD	Multi User Domains
58	MX	Mail Exchanger
59	NCP	Network Control Protocol

60	NIST	National Institute for Standards and Technology
61	NLP	Neuro-Linguistic Programming
62	NNTP	Network News Transfer Protocol
63	NPBS	Norwich and Peterborough Building Society
64	NPL	National Physics Laboratory
65	NSF	National Science Foundation's
66	PAN	Practise Area Network
67	PC	Personal Computer
68	PDF	Portable Document Format
69	PEP	Personal Equity Plans
70	PEST	Political, Economic, Social and Technological
71	PGP	Pretty Good Privacy
72	POP	Post Office Protocol
73	PRNET	Packet Radio Network
74	PS	Packet Switching
75	PSInet	Performance Systems International
76	RBS	Royal Bank of Scotland
77	RIPE	Reseaux IP Europeans
78	TCP	Transmission Control Protocol
79	TPN	Trading Process Network
80	URL	Universal Request Locator
81	UUCP	UNIX to UNIX CoPy
82	VAN	Value Added Network
83	VAR	Value Added Reseller
84	VPN	Virtual Private Network
85	WAIS	Wide Area Information Servers
86	WAN	Wide Area Network
87	WWW	World Wide Web
88	XML	Extensible Markup Language

Chapter 1.

Setting the Scene

1.1 Introduction

The Internet is having a pervasive effect on our lives. Many terrestrial activities in the economy are increasingly migrating to the Internet (Foley and Jayawardhena, 1999). Additionally, a large number of new activities and new entities are constantly being invented on the Internet. For many, the generic term electronic commerce envelops most of the commercial activities on the Internet.

Electronic commerce is a trading innovation that impacts significantly on most academic traditions. For example, it transforms geographical notions of distance and location in relation to retail operations (Bell *et al.*, 1998), it changes the economist's concepts of transactions costs and relative prices (Choi, *et al., 1997)*, and the sociologist must study its contribution to the closure of conventional branch retail outlets, the consequent displacement of certain occupations, and also the social distribution of Internet customers and the possible exclusion of many social groups. As an innovative product and service delivery system, however, it probably has wider implications for marketing than for any other discipline.

Despite the considerable amount of attention given to the Internet in the popular press and the belief that in many business circles that the Web represents a phenomenal marketing opportunity, to date, there has been very little quality academic research on the impact of the Internet on the marketing discipline (Jayawardhena and Foley, 2000).

Electronic services in general, and Internet banking in particular, represent major challenges to the researcher. Firstly, services have unique characteristics that set them apart from physical goods, and the Internet, tends to 'commodify' traditional services (or tum them into quasi physical goods). Secondly, as stated above, much of the marketing literature deals with the terrestrial world, and electronic commerce age literature is scarce. Thirdly, a large proportion of the existing literature deals with the unique characteristics of services and marketing strategies to deal with these, for example, how to measure service quality in services, how to carry out price promotions in services, the impact of advertising on services, etc. By contrast holistic approaches to the study of consumer behaviour, the determination of customer choice and purchase has been relatively under-researched (Gabbot and Hogg, 1998). Finally, perhaps most importantly, the study of Internet banking is of timely importance because;

- banking is a major contributor to the economy,
- traditional banking systems and methods are undergoing significant change with a steady increase in the use of technology, particularly Internet technology, with significant consequences to both consumers and banks, and
- it is a service that every individual consumes almost daily and is becoming ever more important that they make the right choice.

These observations concerning Internet based services, such as banking, give rise to a fundamental question; *what characteristics of an Internet banking service influence consumer purchase intentions?*

As stated above, gaps in the literature and lack of understanding of services delivered over this novel medium necessitate an exploratory approach. Therefore, this thesis draws from the more general literatures on banking industry, consumer behaviour, pricing, branding and service quality. Although less explicitly recognised in the literature, the features of services, including financial services, will also have an important impact on buyer behaviour. Thus any attempt to understand the purchase decision making process must recognise ways in which buyer behaviour can be affected by these characteristics.

Furthermore, since Internet retail banking ostensibly removes most human contact from service provision, it has profound implications for services marketing, and research methodology in particular. Therefore, this research fully exploits prospects presented in these novel circumstances by taking the first step toward understanding these implications for marketing research, by drawing attention to the methodological opportunities offered by Internet banking services. This study moves *away* from the traditional attitudinal research that has been used in previous research, and presents the development of the mystery-shopping research methodology as a novel and appropriate approach to understanding consumer behaviour in Internet banking services.

There are two main of problems with data in Internet banking services, firstly, the lack of audited data, and secondly, the absence of time series data. The combination of these two factors renders full hypotheses testing impractical at this stage. Therefore, the development of the research methodology aspect becomes the most important contribution of this thesis, and the research question provides a foundation and also a means to achieve this objective.

1.2 Definitions

It is traditional to start any analysis by defining terms, it is therefore appropriate at this point to consider the meanings of two of the key concepts in this research, i.e. the *consumer* and *Internet banking services.*

1.2.1 The Consumer

Consumers are often referred to as customers, but in general the term 'customer' is typically used to describe someone who purchases a product from a business. This is a very narrow view since it treats consumer purchase as consisting of only a discrete transaction, i.e. the customer is only of interest in terms of what they buy from the bank. While this is obviously of concern to the individual bank, it is clearly a valid perspective only as much as it describes a small part of a complex set of behaviours. Gabbot and Hogg (1998) assert that there is also a construction on the term 'customer' which implies a simple economic relationship between a business and a buyer, i.e. the relationship is based upon monetary exchange. This view has been broadened over the years to recognise that it is not necessary to engage in any direct financial relationship to consume. Therefore the use of the term 'customer' to characterise consumption is unsatisfactory and other distinctions need to be drawn. What is of interest to this research is the person who makes or influences a purchase decision and this may or may not be the buyer. In a typical household the person who buys banking products may not be the only person who consumes it and similarly the decision of preciously what to buy may be taken in consultation with other members of the household. As a result the term buyer is equally unsatisfactory in describing the full range of consumption behaviour.

The common way to address these ambiguities is to present a definition which encompasses all possible roles played during the consumption process. These include the initiating or authorising role where the persona may determine that some need or want is not being met - the influencing or information generating role of someone who, while not necessarily party to the decision, has a bearing upon the outcome; the buyer who will actually conduct the transaction; and finally the user who is involved in the consumption of the service. Whether all these roles are adopted by a single person or many individuals, these are all facets of consumer purchase behaviour and are, therefore, relevant when considering the consumption of banking services and products. The term consumer is of a higher order and encompasses a number of relationships including the specific cases of the customer and buyer. As a consequence; consumer behaviour defines a wide range of activities and behaviours, the processes involved when individuals select, purchase, use or dispose of products and services. All of these activities have implications for purchase and re-purchase behaviour and can be affected by marketing activity to different degrees.

1.2.2 Internet Banking Services

There is very little academic work that examines Internet banking services (Jayawardhena and Foley, 2000). Within this limited research background, none have attempted to define an Internet banking service. In this context, two subjects need addressing. Firstly, the Internet, secondly, the provision of banking services over the Internet.

The Internet is a network of computers reaching every country in the world. The World Wide Web (WWW or 'Web') began as a very small part of the Internet. The Web is that part of the Internet that can be 'surfed' by following hyperlinks, the underlined text that takes a user from one Web site to another. The interconnections between Web pages evoke the image of a spider web, hence the name.

Most retail banks, cater to the needs of consumers, who are either individual(s) or businesses. Banks satisfy the needs of their customers by delivering a range of banking service products. Traditionally bank branches have served as the distribution channel for banking services for all types of customers. With the advancement and reduction in the cost of technology, the use of technology by banks increased. These brought tangible benefits to customers, among them, the ability to use any branch of the same bank for transactions to the now ubiquitous Automated Teller Machines (ATMs). Therefore it was an incremental progression of the use of technology that made it possible to deliver banking service products remotely, initially by using telephones, followed by dedicated dial up networks (also referred to as PC banking) and now by the Internet. As the opening paragraph stated, in addition to this migration process taking place in traditional banks, new firms have been set up and are offering banking service products over the Internet.

At this stage, it is important to emphasise the distinction between PC banking and Internet banking. Prior to the proliferation of the Internet, the only way that banks could offer electronic banking services to customers was to link customers with the bank electronically through the use of a dedicated dial up network. Essentially this meant that the customers would get connected into a computer network of the bank, through a telephone line or other suitable connection channel. Such a private network was perceived to be more secure than public network (the Internet is a public network) and consequently drew less security concerns. However with the widespread availability of the Internet, installing and maintaining such private network services became disproportionately expensive. Moreover advances in network security technology meant that public network access became progressively safer. Accordingly, virtually all banks that offered PC banking encouraged their users to progressively migrate to Internet banking services.

For the purposes of this thesis, Internet banking services encompass, both traditional banks as well as new bank entrants delivering savings and current account banking services over the Internet. When the term traditional banks are used, they include any traditional institution that has been providing current and savings accounts on a historical basis, for example, Building societies. As of the reference date of this thesis, January 1, 2000, there were fifteen Internet banks operating in the UK. The research limits the examination to personal consumers, and do not examine corporate consumers.

1.3 Background

The banking sector is subject to both internal and external forces (Nellis, 1998 and Rajan, 1998). External factors can be conveniently categorised under political, economic, social and technological changes. In the wider business environment, they are likely to have the greatest impact on the sector. Such developments are, by definition, beyond the control of the businesses themselves.

1.3.1 External Forces

Developments in technology have dominated the revolution in the banking sector during the last decade (Gandy, 1998a). The world-wide expansion in technologies for connection has supported increased globalisation of capital flows and financial organisations. Technology has also facilitated the proliferation of new products and services supporting new consumer demands. Competitive pressures will intensify as organisations seek ever greater productivity and efficiency improvements to sustain profitability.

There has been a marked trend towards deregulation over recent years in many Western countries resulting from political and ideological changes (Nellis, 1998 and Llwellyn, 1996). Successive regimes have continued to deregulate industries as a response to past recessions and to support structural change by improving the efficiency and competitiveness of both public and private sectors. The pace of deregulation has not been uniform across all countries. However, the UK has been at the forefront, with the consequence that the banking sector, in particular, is now one of the most deregulated of all sectors of the economy (Nellis, 1998).

Changes in demographic and social trends have, to a large extent, driven the regulatory and economic shifts that have taken place in recent years (Llwellyn, 1996). These factors will increase in importance over the next decade as the number of young people entering the labour force continues to decline, while an ageing population continues to dominate government welfare spending and the financial necessity, therefore, for greater self reliance. These unavoidable developments have profound implications for the kind of products and services that business will provide and how they are delivered. In this context, the banking sector lies at the forefront of change in terms of society's' needs. At the same time, working patterns will change, with more part-time, temporary and home based workers. This will have further implications for the banking sector.

With the possible exception of regulatory reform, technological change is likely to have the greatest impact on the banking sector over the next decade (Bednar, *et al.*, 1995). Technology is frequently touted as a, if not the, key element in the formulae for productivity and profitability in the 1990s and beyond. It is likely to be the key factor driving change within the banking sector for the foreseeable future.

1.3.2 Internal Forces

The importance of the consumer is at a peak in the banking sector. Legislation has increased consumers' rights, technology and competition has increased their choice of products and providers. The Internet is bringing about changes in the working environment, living conditions and patterns of banking use (Hagel, *et al.*, 1997). These changes will inevitably place consumers under a different set of conditions resulting in changes in their behaviour. One of the outcomes of these changes will be growth in consumers with more sophisticated needs. Consumers will become more discerning as information becomes more accessible over the Internet. Switching between financial institutions and products will continue to grow as consumers' traditional inertia declines. For instance, Internet enabled consumers will be able to change banks at the press of a button, in the comfort of their homes. They will have access to on-line "intelligent agents" that will give them the ability to compare products and services for the best terms and conditions (Rogerson, *et al.*, 1999).

Within the supply chain of the industry, institutions are faced with the challenge of achieving the right balance between staffing levels and skills, investment in technology and branch networks (Gentle, 1993). As organisations seek new sources of revenues and profits outside traditional banking disciplines, they will demand different skills and aptitudes from their staff. This demand, coupled with cost-cutting and the impact of new technology, has already led to a significant reduction in overall staffing levels within the banking industry. Further changes can be expected with the implementation of Internet enabled delivery mechanisms.

As a result of the developments discussed above, the attractiveness of this sector to a wide range of potential new entrants has increased. The cost of entry to the banking sector is low, returns seems very promising and the risk seems manageable. As a result there has been a flurry of new entrants (Gandy, 1998a, Mols, 1998a). Non-bank entrants are exploiting the unique capabilities of electronic networks and leveraging their own resources through Web-based strategies. New technologies, like smart cards and software cryptography are also lowering entry barriers to the banking business, enabling non-banking competitors to take away more and more of the profitable elements of the banking business (Hagel III *et al.*, 1997).

These circumstances present opportunities and challenges to researchers. In times of industry change, research methods examining such industries must suitably be adopted to cope with changes. Opportunities are also abound to present new research methods. The challenge lies in presenting research methodologies that are dynamic and suited for the purpose and yet conform to academic traditions.

1.4 Methods

It is customary to address any research question with a review of previous literature on similar research questions and related subjects. This research centres around literature on,

- financial services in general, banking in particular,
- consumer behaviour in making purchases, services in particular,
- the Internet, and
- market research methodologies.

Financial services sector of many countries, including the UK have undergone many reforms since the 1980s, and yet it has received remarkably little attention from academics and policy makers (Gentle, 1993). In addition, Daniel and Storey (1997) observe that although electronic banking has been available since the early 1980s it is still at an embryonic stage and therefore academic research is needed in this newly emerging delivery channel. To compound the problem academic literature on the Internet and related aspects is rather thin (Day 1997; Chaffey, *et al.*, 2000). This is because, firstly, it is a new area and many people are simply finding their way for the first time, secondly, those who are at the forefront of the Web, the design technologists, are not typically inclined to sit back, reflect on their practise, source relevant theory and write about it. Therefore the methods and approach adopted to address this research question reflects these particular circumstances, and it is important to state the limitations and potential drawbacks.

As the next section of this chapter describes in detail, this thesis begins by examining the characteristics of the banking sector, and thereby presents the state of the present banking environment. The next step is to illustrate how consumers make purchase decisions, initially with regard to both goods and services, followed by services in particular. This narrowing down process continues with an examination of consumer purchases of financial services. Thereafter characteristics of Internet banking services are explored, which enhance the significance of this research for it necessitates the re-appraisal of a service.

The literature review, in the above order, provides the impetus for the generation of working hypothesis. The next step is to devise a methodology to test the hypotheses. At this juncture the complexities of carrying out empirical research in Internet related services are highlighted. While many authors are quick to point out the advantages of the Internet for marketing (Strauss and Frost, 1999; Sterne and Priore, 2000; Sterne, 1996 and 1999; Janal, 2000; Chaffey, 2000), they fall short of proposing of robust marketing research methods that can be successfully employed and at times ignore methodological limitations. Others such as Pincott and Branthwaite (2000) discuss the complexities and difficulties faced in developing new Internet market research methodologies. Hoffman *et al.,* (1997) state that 94 percent of Web users have refused to provide information at a Web site, and moreover of those who do provide information, 40 percent of users give false information. Such findings question the validity of research conclusions arrived at by utilising existing methodologies adopted to the Internet. Many authors reiterate the need to continue refining existing methodologies and to propose new methodologies.

This being the background, a prudent approach is to undertake the process in stages. As a first step, research methodologies adopted in previous empirical work is scrutinized. This procedure reveals restrictions of these methods, and the limitations that will surface if they are adopted for the Internet. In these novel circumstances, this thesis takes a first step towards understanding these implications for marketing research and proposes a new methodology. It proposes a revival of a marketing research methodology that has been rarely used by academics: mystery-shopping.

This is in turn becomes the central contribution of this thesis. Consequently an 'Internet mystery-shopper' research methodology is introduced. The primary data collection instrument in this research comprises of a series of Internet scorecards that makes it possible to carry out detailed cross sectional evaluations of Internet banking service Web sites and associated features. Detailed analysis reveals that, on balance, the proposed methodology is cost effective, flexible and quick compared to attitudinal research methodologies such as those used by previous researchers to examine similar research questions in traditional financial services. The methodology is justified on indicative criteria governing good research design.

Due to the ephemeral nature of Web pages it is essential that, (1) a reference date is assigned for the research, and consequently January 1, 2000 is set down, (2) a copy of each Web page is stored for analysis and subsequent replication. The sample of banks on which the cross sectional analysis is carried out comprises of the entire population of entities providing Internet banking services in the UK as of the reference date. As of this date, there were fifteen Internet banking services operating in the UK. Additionally, as stated earlier, the research is limited to (1) savings and current accounts offered by these Internet banking services and (2) account holders are limited to personal accounts.

As noted by Pincott and Branthwaite (2000), marketing research methodologies proposed at this early stage in the commercial adoption of the Internet cannot be expected to be fully validated, because more empirical work is needed. Consequently the Internet mystery shopping methodology cannot completely test the working hypothesis in this research. As stated earlier, the problem is exacerbated due to the non-availability of audited time series data on customer demand. Therefore what is feasible at this stage to carry out an exploratory analysis and propose possible of improvements that could be made in the future.

The demand for traditional services is heavily subjective primarily because it is a service that is delivered with face-to-face contact. While a number of the evaluation criteria in Internet scorecards used for cross sectional analysis of Internet banking services are subjective, the Internet reduces this aspect and makes genuine mystery shopping as practical as possible. Perhaps the strongest argument in favour of mystery shopping in Internet banking services is that subsequent experience is likely to be identical to a trial, which can be experienced though the browsing process of the web site and observing a demonstration. In most traditional services a trial is not a common option. Although it is possible to try one service episode to establish whether one will purchase a full service or a sequence of episodes, there is no guarantee the future service experiences will be identical.

1.5 The Plan of the Thesis

The chapters in this thesis are organised and presented in the following manner.

A common format is adopted in all chapters. Every chapter, apart from this chapter and the concluding chapter, commences with an introduction that outlines the main areas of consideration, and a chapter summary that highlights concluding remarks and sets out the landscape for the following chapter.

Chapter two examines the characteristics of the banking sector by extending the background discussion. It finds that many of the traditional underlying reasons for financial intermediation are being eroded by related changes in the external environment. While these changes in the banking sector have adversely affected the traditional banks, they have paved the way for the emergence of new entrants. The final section of the chapter expounds the advantages accruing to the banking sector brought about by technology, and in particular the role played by Internet technology in delivering banking service products. This chapter is important in detailing the implications for consumers, in terms of widening choice of banks, increasing ability to bank through multiple channels and the control they can exercise on the services delivered by the banks.

The objective of chapter three is to understand the consumer purchase process. *This* complex task is achieved by dividing the chapter into sections, each section drawing from an interconnected but disparate strand of literature. Consumer behaviour in the purchase process is examined through an analysis of models of buyer behaviour and by investigating the factors affecting purchase intentions of consumers. It finds that the concept of the cognitive consumer is more suitable as a framework for analysis. It suggests that the purchase process is a sequence of tasks characterised as a problem solving exercise. The unique characteristics of services, *intangibility, inseparability, heterogeneity, perishability* and *ownership,* highlights the role of information in the pre purchase stage of services. The importance of service quality, price, brand and risk in the purchase of a service are examined in detail. This examination identifies critical roles of service quality and satisfaction in the formation of consumers purchase intentions. A critique on the exact nature of how the relationship between these two constructs combine to impact consumer purchase intentions finds that empirical efforts to validate the specific nature of the relationship have supported both possible relationships between constructs, i.e. that service quality leads to satisfaction that in turn influence purchases intentions, and that satisfaction leads to better service quality, which in tum influence purchase intentions. The stance taken in this thesis is that service quality judgements form the level of consumer satisfaction which in turn influence purchase intentions.

While the third chapter examines the purchase process in a generalist context, the fourth chapter begin to narrow this process to financial service purchases and finally to Internet banking services. This type of 'focusing' approach is useful because it highlights the fact that consumers face far greater problems when buying financial services, compared to other services, for financial services are often highly complex and may be difficult to evaluate both at pre and post purchase compared to other services. Empirical research work in financial services identifies the effects of service quality, price, perceived value and branding in influencing buyer behaviour. As a prelude to the process of hypothesising the factors influencing purchase intentions in Internet banking services, which is the subject of the next chapter, this chapter examines the characteristics of the Internet and Internet banking services in particular. The analysis reveals that Internet banking service characteristics bear more relation to physical goods as opposed to services goods.

Chapter five generates working hypotheses. The hypotheses are necessary to develop a framework within which the methodology to be constructed, which becomes the central contribution of this thesis. Eleven hypotheses are derived, these hypothesise that purchase intentions are influenced by information, technical and functional qualities, prices, branding, service quality and security of an Internet banking service.

How can these hypotheses be tested? The sixth chapter attempts to address this issue by developing a new research methodology. As is the convention in academic work, the process begins by analysing research setting in traditional banking services. Shortcomings of these existing methodologies are presented, and the case for a new research methodology is put forward. The proposed methodology is drawn from participative observation in the terrestrial world. Although its use in the academic world is limited, mystery shopping, the name given to commercial use of participative observation, has been extensively used in banking in a commercial context. Methodological justifications are made by examining how the proposed research design measures up to criteria governing good research design with respect to measurement, validity and other criteria governing research.

Chapter seven operationalises the mystery shopping methodology. The development of scale measures in the research instrument, the Internet scorecard, is one of the most important aspects of this thesis. The chapter presents four series of Internet scorecards, each dealing with a separate but interrelated aspect of an Internet banking service.

The penultimate chapter achieves two objectives. Firstly, by carrying out an exploratory analysis it demonstrates what can be achieved with the limited amount of data available, and also indicates the potential for the future development of the research. As stated in elsewhere in this introductory chapter, two factors render full hypothesis testing infeasible at this stage, but nevertheless are still valid. Therefore the findings in this exploratory analysis are limited to indicative trends of influences on purchase intentions. Secondly, it offers a glimpse of the factors influencing consumer purchase intentions in Internet banking services from both traditional banks and new entrants, by presenting tentative findings of the mystery shopping research survey. It finds that consumer purchase intentions are influenced by prices, time elapsed since launch, branding, service quality and security of an Internet banking service.

The thesis concludes with a chapter analysing the main research findings and suggesting policy recommendations.

Chapter 2.

Characteristics of the Banking Sector

2.1 Introduction

The introductory chapter presented a brief background of the banking environment. This chapter examines the sector in more detail, objectives being, firstly, to analyse the changes that are taking place within the traditional banking sector, secondly, to explain how these changes are bringing about opportunities to new Internet banking entrants into the sector, thirdly, to analyse how the traditional banks are responding to these changes, and finally how the combined effects of these changes influence consumer choice and by extension their purchase intentions in banking services. A useful and prudent starting point to this analysis is to examine the theoretical underpinning of the need for financial intermediation. This illustrates many of the structural problems that the traditional banking industry is facing, including the threat posed by the emergence of new entrants and the central role of technology, more specifically Internet technology.

The chapter is divided into three sections. It commences by making a succinct examination of the characteristics and change taking place in the banking sector. The next two sections explore issues that appear to be the most important areas of change; the entry of new players to the banking services sector, followed by the role played by the single most important factor that underpins this research. The former highlights the challenges that traditional banks face from these new entrants and the latter expounds on the advantages brought about by technology, Internet technology in particular. The role played by Internet technology in delivery strategies is examined in detail in this section. The chapter concludes with a summary of the chapter.

2.2 Characteristics and Change in the Banking Sector - An Analysis

This section will analyse the inherent characteristics of the traditional banking sector, and investigate how these are changing due to various interrelated factors. In this process it will highlight how these changes are influencing banking organisations with respect to methods of operation and the characteristics of the banking service products on offer. Inevitably, change in the way that banks operate and the banking service products on offer has implications for consumers, more specifically for consumer purchases.

In various ways, related pressures of political, economic, social and technology have eroded some of the comparative advantages of banks in their traditional financial intermediation business. These pressures have brought about more specific changes, which can be better explained conveniently under the following titles.

2.2.1 Monopoly, Comparative Advantage Erosion and Deconstruction

Banks have lost some of their traditional monopolies in many ways. In particular, the development of technology has lowered entry barriers as has the process of deregulation (Nellis, 1998). The process of deconstruction also means that new entrants can offer competition to banks because they are no longer required to provide the full range of banking services, or undertake all of the processes involved in supplying banking services (Evans, 1999). In addition, consumers now have more information about a wider range of alternatives to bank deposits for holding liquid funds. The impact of the Internet is evident in many of these issues, and these are explored in detail below.

New technology and declining entry barriers have also challenged some of the traditional comparative advantages possessed by banks. In particular, disclosure laws have eroded some of the information advantages traditionally held by banks and the development of unit trusts and money market mutual funds also allow consumers to have diversified portfolios even with relatively small investments (Llewellyn, 1996). The development of credit scoring techniques also means that the credit standing of borrowers can be assessed without the necessity of the information derived through an institution maintaining a borrower's current account. For example, cahoot, a new Internet banking service, determines an individual's creditworthiness online within a matter of few minutes. Based on this risk assessment it is able to differentiate among customers by offering them different interest rates and vary the range of facilities offered.

Comparative pressures intensify when competition develops from outside a traditional industry as entry barriers decline (Evans, 1999). This is partly because new entrants often have different cost structures, are less bound by fixed costs, and are often more prepared to challenge traditional industry practices. An Internet banking service can offer banking services to an entire population of a country from a single physical location. Hence, their cost structure is likely to be radically different from a traditional bank with hundreds of branches. Llewellyn (1996) observes that technology is eroding some traditional innocent entry barriers (such as scale factors and the requirement for a branch network for the delivery of financial services), and competition and changes in regulation are eroding some traditional strategic entry barriers (such as restrictive practices, cartels and anti-competitive mechanisms).

A further feature reducing entry barriers is the process of deconstruction (Evans, 1999). This involves the process of decomposing services into their component parts, which may then be provided separately. These need not be undertaken by the same firm, and if, for any reason, different firms have different comparative advantages in different parts of the process, the logical development is for each process to be supplied separately by the firm which has a comparative advantage in doing so. Banks which have an efficient capability for originating and administering loans (because they have a branch network) may not necessarily be the most efficient at holding assets on the balance sheet. Administration and disbursement of loans is becoming a very competitive business. For example, the supermarkets, aided by a very low cost base in distribution (through the existing network of retail outlets and the telephone to some extent) have succeeded in giving loans at very low interest rates. The administration of these loans is contracted out to a bank, for instance the Bank of Scotland (BOS) is administering the Tesco supermarket banking scheme. Another example is the process of securitisation of bank loans: a bank makes a loan, temporarily holds it on the balance sheet, but subsequently securitises it on the capital market. Equally, in some cases the monitoring of borrowers may be undertaken by rating agencies. This is specially so in the case of large companies, whose credit rating is monitored by rating agencies. A lowering of the credit rating is perceived to be an increase in risk, and hence lenders seek correspondingly higher returns when lending to such companies with lower credit rating.

This process of deconstruction or unbundling effectively lowers entry barriers as it means that new organisations are able to enter a market because they need not be involved with the whole process. They can concentrate on that part of a business where they have a comparative advantage. New entrants often target niche markets. This is also related to the question of economies of scale. The major economies of scale in banking relate not to institutions but to processes and functions. For instance, studies suggest that once banks pass a threshold of about $20bn in assets (about one fifteenth the size of Barclays), banks actually get less efficient overall as they get bigger (Graham, 1996). Only a few activities, such as cheque and credit card processing or mortgage lending can be turned into the sort of large operations that produce economies of scale. In general, specialist providers tend to be more efficient than others. As it will be seen in section (3.2), new entrants are the beneficiaries of this process.

One of the major pressures in the banking industry in the years ahead will be the deconstruction process where each institution concentrates on that part of the business and those processes in which it has a comparative and competitive advantage (Llewellyn, 1996). first-e, a new Internet bank, commenced operations in September 1999. It unbundled the banking process by outsourcing the back office operations; the collection and processing of cheques, money deposits, etc. from the Royal Bank of Scotland (RBS); the telephone support centre too was outsourced from a third party. At the outset, first-e was offering only savings accounts. However through this process of unbundling, first-e was able to gain a competitive advantage over traditional banks, offer above market rates of interest on savings accounts, build up market share and began to offer current account services by April 2000.

Similarly, developments in the application of options and asset pricing theory, securitisation, and the evolution of contingent claims and guarantees, have led to an unbundling of the services traditionally provided by banks into their constituent components. Some of these services can now be provided more efficiently in the capital market.

2.2.2 Erosion of Cross-Subsidies

Cross-subsidisation is a common pricing strategy m multi-product firms including banking where, because competitive conditions between different banking markets are not homogeneous, prices of individual 'products' do not accurately reflect relative costs and risks (Martin, 1998a). Cross-subsidies exist between customers, products and processes. This necessarily implies 'subsidising' and 'subsidised' products, which also presupposes an ability to segment markets. As competition intensifies, however, and particularly as economic or regulatory entry barriers are lowered, it is: frequently 'subsidising' markets that are targeted and this erodes the 'excess profits' earned by existing suppliers (Parkin and King, 1995). This in tum forces a change in pricing strategies, which, on the assumption that the cross-subsidisation was designed to raise overall profits, has the effect of eroding aggregate profits. It is partly because banks cross-subsidise parts of their business that new competitors have been able to enter some niche segments of banking business (Gandy, 1998a). However, this entry is also likely to erode the banks' ability to sustain cross-subsidies.

In many countries, banks earn significant endowment profits through 'free resources' (reserves and interest-free deposits) (Gandy, 1998b). These endowment profits have been eroded due to competitive pressures and the deregulation on interest rates. Consequently, a significant traditional source of profits has become less powerful (Nellis, 1998). Historically, the existence of endowment profits due to banks' access to cheap retail funds has acted as an entry barrier to foreign banks (Llewellyn, 1996). To the extent that the cost of retail deposits rises towards the level of wholesale funds, the implicit competitive advantages enjoyed by banks with access to retail funds is eroded and foreign banks and new suppliers are able to compete on less disadvantageous terms.

Cross-subsidies within banks are becoming vulnerable because entry barriers are declining due to the process of deconstruction noted earlier. The general prediction is that, as competition develops (more so in those countries where competition in banking is rather constrained), the potential for banks to engage in cross-subsidising pricing behaviour will be eroded (Rajan, 1998). This would be a: further factor reducing overall profitability. The erosion of cross subsidies has the effect of raising costs on some services, reducing profits, and for reasons associated with endowment profits, can have the effect of lowering entry barriers (Llewellyn, 1996). The debate during much of 2000 covering ATM charges is an interesting case point. Traditional banks own the majority of ATMs in the UK network of ATMs, the Link Network. With the erosion of cross subsidies, traditional banks are beginning to realise that maintaining a network of ATMs has become a significant cost and that they are also at a comparative disadvantage to other banks that are members of the Link Network but do not own a large number of ATMs. This has given way to the threat from a number of ATM owning banks that the cost of usage will be recovered from non-customers: i.e. if bank A owns a large number of ATMs, and a customer from bank B uses an ATM belonging to bank A, the customer faces a penalty charge. This issue has attracted attention from banking commission appointed by the government to look into the state of banking affairs in the UK.

2.2.3 Locational Factors

Traditional banking is characterised by physical decentralisation, with branches scattered around populated areas providing an ubiquitous presence. The rationale behind such branch investment is the need to distribute banking services. A high street presence encourages usage, maintains contact with customers and provides the necessary visibility to the general public. A large branch network provides customers, among other things, the freedom of carrying out their banking activities devoid of geographical limitations, the reassurance that the bank has substantial resources and hence offers security for their savings. This has enabled large traditional banks to attract customers and continue to maintain their market share. Therefore the spatial element has been central to the delivery strategies of traditional banks. Such a structure allowed these institutions to provide banking service products at the cost of a large number of staff with high fixed and variable costs. In order to cope with fluctuations in demand on each location, more resources are employed resulting in sub-optimal capacity utilisation (Gentle, 1993).

The substantial cost and effort to develop such a network has often proved to be an entry barrier to prospective competitors, and helped to maintain the concentration rates and the oligopolistic nature of banking in the UK (Henderson, 1995). The position of the banks is only tenable, so long as they are not superseded by more cost-effective or efficient delivery systems, and provided they meet the locational needs of their customers (Gentle, 1993). Clearly, entry barriers arising from these factors are declining now with the large-scale adoption of the telephone, and more recently the computer, for the delivery of banking products.

The significance of locational factors are illustrated in the case of first-e. first-e, has acquired it's banking license from France, has a registered office in Paris, France, but operates in the UK. Under the European Union rules it is possible for an organisation with a banking license from one of the member countries to operate throughout the union. The bank's telephone support centre, outsourced from a third party, is based in Dublin, Ireland. Royal Bank of Scotland handles first-e's banking back office functions. The corresponding office is situated in Milton Keynes, UK. With this locational spread over several countries, first-e demonstrates the changing nature of locational factors.

2.2.4 Own Margin

If banks' supply price of financial intermediation rises (as measured by the interest margin) banks may become relatively less competitive (Gentle, 1993). This may be because, as explained above in sections (2.2.1) and (2.2.3), they are locked into a traditional cost structure due partly to having invested substantially in a branch network that is no longer the only means of delivering financial services. At the same time, the cost of capital (relative to the cost of debt) has in general risen for banks, partly because their risk profile has tended to deteriorate somewhat due to competitive pressures (Gandy, 1998b). Further, as previously noted, competitive pressures have eroded the ability of banks to engage in cross-subsidy pricing. This in tum implies that previously subsidised parts of the business are less viable, to the extent that banks are forced through competition to lower the price components of the business.

For similar reasons, competition has reduced the endowment profits so much so that competition is forcing banks to pay a rate of interest on a higher proportion of deposits at a rate close to market levels (Llewellyn, 1996). New Internet banking service entrants such as, egg and first-e, are pushing margins to unsustainable levels by offering interest rates on savings accounts over and above the Bank of England base rate. Moreover, for a period of time egg offered a mortgage interest rate below that of their own savings account rate. Clearly such a 'loss-leader' strategy is unsustainable in the long run and is designed to capture market share. Nevertheless such dramatic offers disturb the entire lending and saving market.

The power of competition is evident from the pressure on banks throughout the world to cut costs, through a combination of reducing the numbers employed and by closing branches. Banks seem to be under considerable pressure to lower the supply price of financial intermediation and to narrow the lending margin (Henderson, 1995).

2.2.5 Lower Costs of Alternative Suppliers

For the same reason, if the costs of alternative suppliers of traditional banking services fall relative to those of the banks, they become potentially vulnerable. In particular, financial innovation and the power of new technology has tended to increase the relative competitiveness of the capital market in relation to banks. Further, new delivery technologies has lowered the cost of alternative suppliers of financial services to the extent that they no longer need to develop a full branch network. The examples highlighted above further augment the constant pressures that traditional banks have to endure.

2.2.6 Regulation

Financial market regulation is arguably needed in order to mitigate the undesired effects of externalities, asymmetric information and monopoly powers (McKenzie and Khalidi, 1996). During the last two decades of the previous century, the banking sector experienced many changes in the regulatory regime under which it is governed. Of particular importance is the fact that this process has been globally distributed. As a result of the convergence of the regulatory regimes of many countries, national borders have become less relevant, and this has paved the way for the globalisation of the financial services sector. Additionally, the internal barriers within the sector that served to distinguish traditional institutions, for instance insurance companies, from banks have become less distinct with de-regulation.

In most instances regulation has the potential to create and sustain economic rents and protection (Coyne and Dye, 1998). This protection frequently leads to increased costs, buoyant profits, and excess capacity. Historically, regulation in banking has been protective and has often had the effect of limiting balance sheet growth and the allowable range of business that banks can undertake. It has also had the effect of limiting competition on the premise that "excessive competition" in banking can lead to increased risk and potential systematic hazards (Llewellyn, 1996). Regulation in banking has often condoned restrictive practices and anti-competitive devices, and has in general had the effect of limiting price competition. As a consequence, profits in the sector have been high. During the last five years, each of the big four traditional banks in the UK, comprising of Barclays, NatWest (now merged with RBS), Lloyds (now merged with TSB) and Midland (HSBC) have on average, consistently earned profits in excess of £1 billion per year. Additionally, the high value of a banking franchise and the low risks emanating from various forms of credit rationing have helped to produce healthy profits. At the same time, costs tended to rise to exploit the economic rents created by the protected environment, and non-price competition (for example the provision of large number of branches within a specific area) has dominated for a long period (Gentle, 1993). This in tum has created an excessive cost structure.

The combined effects of these have resulted in excess capacity. This capacity in the banking industry is viable so long as it is protected, but will prove to be unsustainable in the absence of such protection. Further, to the extent that regulatory costs imposed on banks are higher than those imposed on alternative suppliers of some of the services provided by banks. Regulation has the effect of increasing the relative competitiveness of non-bank suppliers of traditional banking services (Gandy, 1998a).

2.2.7 Excess Capacity

Since entry barriers are declining faster than exit barriers, it is likely that excess capacity will emerge in the industry. If new entrants believe they have a competitive advantage in relation to existing players, they will enter the sector. This will create excess capacity, and yet more entrants will continue to enter in spite of excess capacity. This results in putting more pressure on existing players. The manner in which excess capacity is removed from the banking industry will be one of the major strategic issues that banks will face in this decade. As it has been observed during the last few years, capacity reduction could take many forms, including, merger of different banks, closure of branches, reduction in the range of product on offer, changes in terms and conditions of banking products, etc. Inevitably, such changes also influence consumers, and this may be more pronounced when consumers are buying new banking services. Therefore these changes influence consumer purchase intentions.

Compared with other industries, the concept of "excess capacity" is more difficult to define and measure in banking. Llewellyn (1996) identifies four alternative concepts.

2.2.7.1 *Excess Capital*

It is very likely that there is an excessive volume of capital in the global banking industry in that, given the market conditions, it is unlikely that the required rate of return on capital can be earned in the long run (Gentle, 1993). Perhaps the market is not big enough to support the current volume of capital in the banking industry. The excesses in capital could be due to two reasons (Llewellyn, 1996). Firstly, regulation imposes an unsustainable capital requirement. Secondly, the business environment has changed in a way that the industry with its current structure and the amount of business it is able to conduct, can no longer support current capital levels. This may be because new firms have entered or because demand has shifted away from banks (for example, business customers switching to the capital market).

Excess capital raises the required rate of return on assets in order to service the capital base. However, the same competitive conditions that have caused banks to lose some lending business also make it difficult to increase the rate of retum on assets. Faced with excess capital, a bank has three broad options (Lumby, 1994; Jayawardhena, 1997). Firstly, it can expand the balance sheet perhaps by making more risky loans which may have the effect of eroding lending margins. However, if this induces banks to make loans without incorporating the true risk premium, in the end there will be a destruction of capital. Secondly, it can make an acquisition. There is ample empirical evidence that banks with excess capital often pay a premium when making acquisitions and this makes it difficult to subsequently earn a sufficient risk-adjusted rate of return on the investment. This process has been put in motion by a number of banks, examples of such acquisitions during 2000 were: RBS acquisition of NatWest and LloydsTSB's acquisition of Scottish Widows. And finally, it can repay capital to shareholders.

The last option may be the optimum strategy if regulation limits the extent to which bank capital can be deployed in new business areas which does not, of course, limit where shareholders can invest externally to the bank. Shareholders have more options to allocate capital externally than banks have internally (Brealey and Myers, 1984):

It is possible to have global excess capital in banking even when each individual bank believes it is short of capital. The two are not contradictory. If each individual bank is seeking to increase its share of a declining market, its own capital may be insufficient to support its planned business profile. But in aggregate, banks may have too much capital for the available amount of profitable business.

2.2.7.2 *Too Many Banks*

It is also evident that there are too many banks that prevent the exploitation of economies of scale (Llewellyn, 1996; Henderson, 1995). Although the empirical evidence with respect to economies of scale in bank firms is inconclusive, there are clear economies of scale in bank processes (Llewellyn, 1996). Banks may merge in order to secure these economies. It is almost certain that there are economies of scale that can be reaped which are being denied by the current structure of the banking industry in many countries. The take over of NatWest by the RBS in mid 2000 is a prime example of this process.

2.2.7.3 *Excessive Infrastructure*

This relates to the infrastructure as opposed to the number of banks. In many countries the number of branches is excessive with an implicit unnecessary duplication of banking infrastructure - fixed costs and delivery facilities. This excess capacity can be reduced either by individual banks closing their branches or by merging banks and closing overlapping branches. The large scale closure of branches in late 1980s and 1990s in the UK was an attempt by the banks to address this problem (Gentle, 1993). In 1990 there were 12,994 branches in the UK, and this dropped to 11,000 by 1999 (Bachelor, 2000).

The process is still continuing, and the closure of 171 branches by Barclays bank in mid 2000 is a good case in point. In effect, a co-ordinated strategy can be more effective than all banks acting unilaterally given that, in some cases, a major benefit from a branch closure can accrue to a competitor that is able to absorb a lost customer base without adding to its own costs. Many bank mergers have been motivated by a strategy of reducing the infrastructure of the combined bank. An alternative strategy when faced with excess distribution capacity is to attempt to supply more products and services through it. Faced with excess distribution capacity, banks have two broad strategic alternatives, reduce capacity or pass more business through existing capacity. In this respect, there is a close parallel between excess capacity in capital and infrastructure.

In this context, technology may prove to do more harm than good. Given that, technology lowers the cost of delivery and lowers entry barriers to new entrants, the prospect of excessive non-terrestrial capacity building by the banking industry is a possibility.

2.2.7.4 *Technological Capacity*

Developments in technology have themselves impacted on capacity in that new technology vastly increases the capacity of banks to supply services. It is unlikely that, given the economies of scale in new technology, the current number of banks can be sustained as they cannot all apply new technology to its most economic extent. And yet banks individually will attempt to do so. This is a case of the fallacy of composition, i.e. what is viable for an individual bank is not necessarily so for all banks taken together (Llewellyn, 1996).

2.2.8 Customer Retention

Traditionally, banks, however much they tried, were hard pressed to persuade consumers to move in any numbers (Gentle, 1993). Bank customers, unlike retail customers, tend to be loyal. Provided they are satisfied with the service, they remain with the same bank for five to seven years on average, and switch only when they move to a new home in an area outside their bank's network (Gentle, 1993).

This explains why banks' market shares in the past have been so static. Assuming that most banks replace customers moving out of the market with people moving in (and replace most of those who choose to leave with new customers), the only extra customers come from net household growth, which seldom rises above 3 percent a year (Huber *et al.,* 1998). Even if a bank was to capture a disproportionate share of this growth, for example say 40 percent, it would be unable to increase its market share by more than one percentage point a year. In the absence of acquisitions, the market share of any one institution rarely grows by more than one or two percentage points a year (Huber *et al.,* 1998).

In the recent past there have been significant changes in consumer attitudes towards banking. Therefore, if customers value the services offered by banks less than in the past, or their preferences shift to alternative suppliers, banks again become vulnerable (Llewellyn, 1996). Legislation has increased customers' rights, technology and competition has increased their choice of products and providers. It may be, for instance, that the recent poor performance of banks in many countries may have eroded some of the reputation advantages traditionally possessed by banks (Gandy, 1998c). Borrowers may also choose to have a more diversified structure of debt and to become less dependent on banks for the supply of credit. As financial markets have broadened and deepened, markets increasingly offer a wider choice of facilities than has been the case in the past.

The Internet will bring about changes in the working environment, living conditions and patterns of banking (Hagel, *et al.*, 1997). These changes will inevitably place users under a different set of conditions resulting in changes in their behaviour. One of the outcomes of these changes will be growth in users with more sophisticated needs. Customers will become more discerning as information becomes more accessible over the Internet. Switching between financial institutions and products will continue to grow as customers' traditional inertia declines. For instance, Internet enabled consumers will be able to change banks at the press of a button, in the comfort of their homes. They will have access to on-line "intelligent agents" that will give them the ability to compare products and services for the best terms and conditions (Rogerson, *et al.*, 1999). This means that customer retention is becoming more difficult.

2.2.9 Asymmetric Competition

Competition has a powerful impact on any industry. To some extent, competition works asymmetrically in the banking industry (Gandy, 1998b). Developments in technology, and the general erosion of entry barriers into banking, mean that it is easier for non-bank financial institutions and non-financial institutions to diversify into banking than it is for banks to diversify out of financial services. For example, the major supermarkets, Tesco and Sainsbury's, offer a range of banking services, but major banks have not ventured out into retailing.

As entry and regulatory barriers are eroded, banks are likely to face competition from a wider range of competitors. In addition to the examples illustrated earlier, several other examples in many countries can be cited where new entrants have been able to compete with banks in supplying some traditional banking services. In-house banks such as Volvo in Sweden, British Petroleum in the UK, Renault in France, have all been able to internalise some of their banking operations and, to some extent, provide a limited range of banking services to others (Llewellyn, 1996). Some large corporate customers have become more credit-worthy, and have higher credit ratings, than their bankers. In these circumstances it is not surprising that they both displace banks and to some extent offer banking services to others. Two of the largest corporate lenders in the USA are the General Electric Company and the Ford Motor Company (Llewellyn,

1996). In some countries, car manufacturers have acquired their own bank for the provision of credit to sales agents. In the USA, industrial and transportation companies, manufacturers and retailers have acquired insurance companies, finance companies and leasing operations. General Motors and IBM offer short term money market facilities and commercial loans to companies. There is evidence that these companies are offering banking facilities to individual consumers (Llewellyn, 1996). In most cases these are offered in conjunction with some product that these companies trade in.

2.2.10 Disintermediation

The much wider availability of low-cost time, or near-real time, information in particular has spurred the process of disintermediation. In theory at least, easy access to information shifts the balance of power towards the end user; if all information is only a mouse click away, people will gravitate towards low cost or best value (Taylor, 1998).

The banking industry was among the first wave of industries to feel the effects of this displacement of intermediaries (Hagel, *et al.*, 1997). As information technology infiltrated the banking industry, and as the industry was deregulated, consumers seemed less and less in need of banks: at least not banks as bureaucratic intermediaries. Initially, bankers saw that many of the business segments they dominated, such as large corporate lending and credit card processing recede from their control. Now this process has spread into individual consumer related banking services with the advancement of Internet banking services. Nevertheless some banks have positioned themselves to the disintermediation threat from Internet banking and bill processing and are seeking to carve out a role as trusted third party, certifying agent, privacy broker for electronic commerce, and last but not least by offering Internet banking services themselves. Barclays Online, the Internet banking service of Barclays bank, is one of the biggest Internet banking services in the UK. As stated earlier, RBS, provides back office operations for a number of new entrants. This is evidence of the traditional banks responding to the changing environment.

2.2.11 The Value of Relationships

Many believe that a competitive environment may threaten relationships. Customers might be tempted to switch banks, depending upon which bank offers them the best offer. Increased credit market competition then imposes constraints on the ability of borrowers and lenders to intertemporally share surplus (Petersen and Rajan, 1995). When parties anticipate a shorter expected "life-span" of their relationships they may respond by reducing their relationship-specific investments. More specifically, anticipated shorter relationships inhibit the re-usability of information, and thus diminish the value of information. Banks may then find it less worthwhile to acquire (potentially costly) proprietary information, and consequently relationships suffer. Paradoxically, shorter or weaker relationships actually become a self-fulfilling prophecy (Llewellyn, 1996).

These arguments highlight the negative spiral that may undermine banking. An important observation is that this negative spiral might be self-inflicted. While competitive banking challenges relationships, the bankers' response of cutting back on information acquisition may actually damage banking relationships most. Relationships facilitate a continuous flow of information between customer and bank which may guarantee a smooth running of accounts. These relationships may give banks a comparative advantage. But also borrowers need to invest in relationships as banking is a mutual commitment (Gandy and Brierly, 1997). Customers, however, face an equal challenge: how to benefit from competitive pricing without jeopardizing the benefits of relationships (Rajan, 1998). This is the *relationship puzzle.*

The relationship puzzle has no obvious solution. Relationships may foster the exchange of information, but may simultaneously give lenders an information monopoly and undermine competitive pricing. The informational monopoly on the "inside" lender's side may be smaller if a borrower engages in multiple banking relationships. This would mitigate the possibilities of rent extraction by informed lenders and induce more competitive pricing (Sharpe, 1990).

Transaction-oriented finance, however, may give little incentive to acquire information, but is potentially subjected to more competition. There might be no winners in this process, for example, transaction-oriented finance may not *be feasible where* relationship oriented finance retreats. More specifically, markets for transaction-oriented finance may fail when problems of asymmetric information are insurmountable. The distinction between relationship-oriented finance and transaction-oriented finance, or between bank-dominated systems and market-oriented systems, may therefore be less well defined than it appears (Gentle, 1993). What might be true is that a bank-dominated system invites oligopolistic behaviour, such that competition is contained (and relationships preserved) while a market-dominated system suppresses competition less. A less competitive financial system may thus *preserve* relationships more. Competition threatens relationships, but it may simultaneously elevate the importance of relationships as a distinct competitive edge. This is the *relationship paradox* (Rajan, 1998).

As it will be seen in chapter four, many Internet banking services shows characteristics of a commodity with apparent low emphasis on relationships. While the acquisition of information in Internet banking services may be cheaper to terrestrial banking, the relationship puzzle is still prevalent since consumer purchase intentions may be strongly influenced by competitor pricing. On the other hand, a relationship orientation can alleviate competitive pressures. Thus, a more competitive environment should encourage banks to become consumer driven, and customise services. Since a relationship orientation may earn banks a substantial added value, banks would then isolate themselves from pure price competition. This is in effect what some of the new entrant Internet banking services are attempting. egg, initially offered an interest rate for savings accounts that was well above market rates. This attracted a large number of customers. When other new Internet bank entrants began to offer superior interest rates, egg did not attempt to compete. Instead it began to create a financial and insurance services supermarket, and attempted to sell these products to customers based on the relationships fostered with them.

2.2.12 The Erosion of the Institutional Stature of Banks

Being a traditional institution gives banks a privileged position or stature, because individuals like to trust their financial affairs to an organization which they believe will behave in ways that will not be detrimental if unforeseen circumstances should arise (Lane and Bachman, 1996; Bidault and Nihtila, 1999). For generations banks have skilfully been able to commercialise trust by differentiating themselves through their institutional stature. Banks serve fundamental needs both an individual level and at a community level, which has far sustained their institutional stature (Lowe and Kuusisto, 1999). The shifting of banks from being a trusted institution of just another company with focus on short-term profits is effectively an erosion of the institutional stature.

Institutional stature can also be a burden for banks because they are not necessarily good at marketing and their attitudes and organisations may not be capable of meeting the challenges of rapidly changing markets (Lowe and Kuusisto, 1999). The erosion can be a slow process and has the potential to creep up on banks without them always fully realising it themselves and conventional measures of customer loyalty can be misleading. In their research, Lowe and Kuusisto (1999), found that the rate at which customers move to other banks was less than 5%. However they also found that an important feature of this apparently very low defection rate which banks may underestimate, i.e. that many customers who switch to new banks leave behind dormant accounts. These dead accounts hide the number of deserting customers and thus maintain a misleading picture of the stability of bank's customer base. These findings signify the level of erosion of the institutional stature of banks.

It was earlier described how technology, Internet technology in particular, makes the transfer of accounts from one banking institution to another, very quick and simple. As noted above and described in detail in section (4.5.4), the commodity like characteristics of an Internet bank account relegates the importance attached to the stature of the bank, with consumers relying more and more on price and account features for purchases.

This discussion has so far illustrated that many changes are taking place in the traditional banking sector due to various interrelated factors. It also demonstrates that a number of factors give rise to falling entry barriers in the banking sector, paving the way for new Internet banking entrants. Further, these changes are influencing banking organisations, methods of operation and the characteristics of the banking service products on offer. Inevitably, such changes in the way that banks operate and the banking service products on offer, has implications for consumers, more specifically for consumer purchases.

What has been evident throughout the analysis is the central the role played by new Internet bank entrants and technology. The following two sections will examine these two aspects in further detail, for they are very important to the central research question of this thesis.

2.3 The Emergence of New Entrants in the Banking Sector

Structural change in the banking sector combined with technological change create an atmosphere conducive to the entry of new entities into the banking sector. Many of these new entrants are utilising Internet technologies for many reasons. Therefore this section will firstly examine the characteristics of new Internet bank entrants, and secondly examine the impact that these new Internet bank entrants are having on the banking services sector as a whole with particular emphasis on how they may influence consumer purchase intentions.

2.3.1 Characteristics of New Internet Bank Entrants

In general, new Internet bank entrants have tended to concentrate on niche markets and have not attempted to offer all intermediation and payments functions offered by banks. For example, egg and first-e, have concentrated on capturing the savings account market share, at least in their formative stages. New Internet bank entrants are able to choose those parts of the business of banks in which they judge they have a competitive advantage and, unlike their traditional banking competitors, are not under pressure to offer the full range of banking services. This occurrence, as described earlier, is partly associated with the traditional banks' pricing policies which in some cases create cross-subsidies, whereby uneconomic services are sustained in part by subsidies from other parts of the business. The new Internet bank entrant, on the other hand, is able to price services without cross-subsidies as they may choose to offer products that traditional banks are over-pricing (Llewellyn, 1996). At present it appears that new Internet bank entrants are taking a slightly different version of this strategy. They are focusing more on acquiring a large customer base, by offering 'loss-leading' products, for example savings accounts with interest rates above the Bank of England base rate, which may influence consumer purchase intentions significantly.

New Internet bank entrants tend also to be highly focused in terms of the value chain. They unbundle much of the high-cost processing, and subcontract these activities to specialists including, in some cases, to traditional bank competitors. The new Internet bank entrant is able to concentrate on that part of the value chain where they have a comparative advantage. Because they have the ability to subcontract large segments of processing, and effectively buy into economies of scale through specialist processes their competitiveness is enhanced. New firms are also often able to operate with comparatively low fixed costs that further lower entry barriers and also allow a high degree of flexibility. As stated earlier, first-e, a new Internet bank entrant is a very good example of this process, its back office operations of money collection and clearing processes are outsourced from RBS, a traditional bank, its call centre support is subcontracted to a third party.

A new Internet bank entrant also avoids the costs of "inheritance". They do not have an existing cost structure based on past technology and uncompetitive conditions. They are able to avoid the transaction costs of re-engineering the business to adapt to current conditions. In particular, it does not have the inheritance of a costly branch structure dedicated to the supply of traditional banking and other financial services.

For traditional banks, devising viable competitive strategies is particularly challenging when it is not clear or certain who future competitors will be (because the economies and competitive strategies of new Internet bank entrants are difficult to fathom as they are different from incumbents). Additionally the process is made even more difficult in instances when new Internet bank entrants are competing in a business which is subsidiary to their mainstream but which is a core business of incumbents. This latter consideration may have the effect of raising the contestability of sector or market segment. Because the activity is not a core business of a new Internet bank entrant, exit barriers may be low. Exit can be undertaken without fundamentally changing the nature of the business (Gandy, 1996). Thus, some of the new Internet bank entrants are competing in business which is core to banks but peripheral to themselves.

The implied increased contestability of banking markets poses a competitive threat to traditional banks (Llewellyn, 1996). New firms may enter banking markets but also have the capacity to subsequently exit the market at low cost. This may mean that there is a constant inflow and outflow of new competitors. While any new comers may not be a permanent competitor there may be a constant inflow of (albeit temporary) new competitors. The range of competitors may be constantly changing as each new Internet bank entrant is experimenting with banking business and it may or may not prove to be permanently profitable to it. Such "trial and error" type of competition offers permanently higher competition to existing players even though the population of competitors may be in constant change. Therefore low exit barriers for new Internet bank entrants pose as substantial a competitive threat as does low entry barriers. Therefore the impact of new Internet bank entrants to banking markets should not be measured in terms of volume or the market share they may gain. The latter may be quite small but the impact on traditional banks could be large because it forces them to respond by offering services at lower cost and/or higher quality. The impact may be substantial because the range of new suppliers of financial services increases the contestability of banking industry. While a new Internet bank entrant is yet to 'exit' the banking services sector, a good example of a new entrant exiting the financial services sector that can be quoted is that of AT&T in the US selling its leasing and financial business in 1996 (Llewellyn, 1996).

While the focus in this section has been on the characteristics of new Internet bank entrants, the characteristics of Internet banking services that both new internet banks entrants and traditional banks offer are examined in the fourth chapter.

2.3.2 Impact of New Internet Bank Entrants

The impact of new competitors is likely to be considerably greater than indicated by their market share. It is unlikely that customers will close existing traditional bank accounts in order to shift business to new Internet bank entrants. This is because new Internet bank entrants usually do not offer the full range of banking services. The impact on traditional banks depend not so much from market share considerations but through the impact on competitive conditions. Banking markets have, as a result, become more contestable and this, rather than market share, has an impact on the banks competitive behaviour. Additionally, as argued above, if entry and *exit* barriers are *low,* there will be a possibility of a constantly changing group of new competitors none of which individually gains substantial market share but which collectively have a significant impact on competitive conditions in the market.

Three developments can be identified in the entry of non-financial companies to banking markets. Firstly, because of increased competition, the value of existing franchises of some of the new Internet bank entrants has tended to decline. Secondly, diversification is often a strategy designed to extend the range of a franchise in the context of a decline in its value in existing markets. In the case of supermarkets, for instance, they have become largely homogenised and it has become increasingly difficult for them individually to differentiate themselves from their competitors. If this is one of the spurs of diversification, the development of technology and deconstruction makes entry into banking more feasible and economic. Thirdly, a parallel decline in the value of the banking franchise creates an incentive for traditional banks to seek new business and this may be achieved through developing a partnership with other companies. This could be construed as a strategy of sharing respective core competencies.

Overall the entry of supermarkets and other organisations into banking illustrates how declining entry and exit barriers can have a significant impact on competitive conditions in banking markets. The supermarkets have come to view themselves not as food stores but as retailers, it is largely in this area that their core competencies lie and existing skills may be transferable to other areas such as banking services.

New Internet bank entrants are able to compete with traditional banks on price, because traditionalists have to spend considerable amounts on integrating its Web site with 'legacy' systems. Legacy systems are any particular processing systems that are carried over from previous generation of technology or methods of working. Legacy systems may be problematic in two ways: disjointed systems and the way data is processed. A bank maintains different processing programmes to deal with different types of transactions, and may separate them by geography as well. Therefore, in practice, there could be dozens of separate programmes running independently of each other and this creates an atmosphere where it becomes very difficult and costly to compete with new Internet bank entrants, who do not suffer from such problems. Related to this, and as previously noted above in section (2.2.1) under deconstruction, the smaller size of new Internet bank entrants places them at a considerable advantage compared to the large traditional banks.

2.4 The Central Role of Technology in the Banking Sector

The interrelationship between technology and the banking sector has been highlighted very prominently in the preceding sections of this chapter. As explained previously in the introductory chapter, Internet banking services is a process where banks harness the capabilities of technology. This section examines the use of technology in the banking sector by paying particular attention to the use of technology in the delivery of banking services. This discussion is important, since the use of Internet technology in banking has important consequences for consumers, with respect to the way consumers purchase banking services and subsequently how they maintain their banking relationships. However, these consumer implications are examined in detail in chapter four and five.

Banks arose historically to provide customers liquidity and a safe investment heaven (Rajan, 1998). Llewellyn (1996) cites that, (1) they may have a particular expertise enabling them to do what other firms cannot do: they possess certain monopoly powers, and (2) they do what technically others can do but they possess certain comparative advantages which give them a comparative advantage in the market place, as being more specific reasons for their existence.

It follows from this that any firm becomes potentially vulnerable if it loses a monopoly power (i.e. others become able to do what was previously the exclusive preserve of the firm(s) in question), or its comparative advantages are eroded. In some areas of business, banks have historically had monopoly powers and comparative advantages. However, as the second section of this chapter demonstrated, both these reasons are now under question in that there is now virtually nothing a bank does which could not be done by markets, non-bank financial firms, or non-financial banking firms. What is also clear is that, in most cases, these changes have been brought about by developments in technology.

One of the major advantages of technology is to make expansion profitable as well as feasible. It does so by consolidating, streamlining and organising back and front office functions in ways that makes sense both for the institution and for the environment in which it plans to expand. Arguably, advances in technology, more specifically advances in mediums of communication and data transfer (the Internet is the most advanced, the cheapest and the most resilient form of this medium), combined with the de-regulation of the financial services markets around the globe have increased the pace of globalisation.

The power of technology will be, and has been, particularly decisive. As recognized earlier in the discussion, technology acts as both a threat and an opportunity to banks. It enables existing services to be provided more efficiently, it enables new services to be offered, it lowers entry barriers in some areas, and changes the economics of delivery. Technology has the power to transform the fundamental economics of any industry and banking is no different from others. It has the potential to increase the availability of information and reduce the cost of access to information. This is a potentially powerful force as it both reinforces and challenges one of the banks' major core competencies - information. Given that banks are ultimately in the "information business", anything that impacts on the availability, cost and management of information must have a decisive influence on their business (Jayawardhena and Foley, 2000). A combination of new technology, the increasing role and power of rating agencies, and more extensive disclosure laws are eroding some of the traditional banks' traditional information advantages. In some cases, information that was previously a private advantage to banks has become more of a public good (Jayawardhena and Foley, 2000).

2.4.1 Delivery Strategies

Traditional banking is characterised by physical decentralisation, with branches scattered around populated areas providing an ubiquitous presence. The rationale behind such branch investment is the need to distribute banking services. High Street presence encourages usage and maintains contact. A large branch network gives customers easier geographic access and the reassurance that the bank has substantial resources and hence offers security for their savings (Lockett and Littler, 1997). The extensive branch network has enabled banks to capture market share. Therefore the spatial element is central to current delivery strategies. Such a structure allows these institutions to provide homogeneous products and services at the cost of a large number of staff with high fixed and variable costs. In order to cope with fluctuations in demand on each location, more resources are employed, resulting in sub optimal capacity utilisation and excess staff. The vast cost and effort to develop such a network has often proved to be an entry-barrier to prospective competitors, and it has helped to maintain the concentration rates and the oligopolistic nature of banking in the UK (Henderson, 1995).

The position of banks is only tenable so long as they are not superseded by more cost-effective or efficient delivery systems, and provided they meet the locational needs of their customers. With the widespread use and adoption of the Internet, there exists the possibility for existing financial institutions and other new entrants to provide their products and services over the Internet. Therefore it is pertinent to explore the advantages that would accrue to banks in utilising the Internet for the delivery of products, services and other uses.

2.4.1.1 The Rationale for Banks using the Internet

The properties of the Internet make it an ideal medium for delivery of banking products and services. The advantages accruing to a bank can be outlined as follows.

1. Cost savings - Internet delivery is cheaper than physical channels (Jayawardhena and Foley, 2000). A simple transaction cost for a non-cash payment at a branch is likely to cost the bank as much as eleven times more than over the Internet. As with all forecasts and estimates related to the Internet, different authors provide disparate accounts of the magnitude of savings. This estimate is based on studies by Downes and Mui (1998), Wylie (1999) and Economist (1999). The cost savings come about through the combined effects of reduction and better utilisation of the workforce, equipment, more economic usage of space and operational savings.
2. Increased customer base - One of the primary objectives of developing new products and services is to attract new consumers and to retain existing customers (Read, 1998). Present Internet demographics suggest that it is the relatively well off and the well educated that use the Internet, which suggests that potential users are high net worth customers (Malaga, *et al.*, 2000). Banks unable to respond to requests for new services risk loosing existing customers to competitors. The use of multiple distribution channels can increase effective market coverage by enabling different products to be targeted at different demographic segments. Additionally, customers are likely to place their trust in proven innovators. Therefore it is important to build a reputation for innovation. This may make it easier to sell financial services, attract more customers, and retain existing customers.
3. Mass customisation - Internet delivery has the capability to customise information to suit the needs and the likes of each user (Dannesnberg and Kellner, 1998). Mass Customisation refers to the notion that each individual user perceiving that the service they receive is personalised or customised to their needs and uses. Such features are becoming more and more important in a world saturated with mass automation and homogenised products and services.
4. Marketing and communication - The World Wide Web has the capability to host advertisements and other marketing campaigns

without an incremental cost structure for prolonged exposure (or display) like those found in the traditional media (Quelch and Klein, 1996). Once a Web page is designed and hosted on a server it serves its purpose 24 hours a day as long as it is active on the server. Costs are limited to initial development costs and maintenance costs that are less in comparison to traditional media. Additionally the interactive nature of the Internet facilitates a system whereby a customer can be guided through a catalogue of products and services that is most suited for them depending on their socio-economic profile (Jayawardhena and Foley, 2000). It is cheaper than traditional mail-shots and far more effective since it is customised. Further, the Internet can be used very effectively to collect customer data with minimum effort. Achieving more operational usage from customer databases is also made easier.

5. Innovation - Internet technologies have paved the way to a multitude of different banking products to be innovated (Prescott and Van Slyke, 1997; Mandeville, 1998). It also facilitates the delivery of products and services in an innovative manner to customers.
6. Development of non-core business - Recent changes m the regulatory framework have enabled many banks to expand their services into non-traditional banking areas (Marshall, 1998). For instance, many banks have already moved into, or are in the process of moving into, insurance and stock brokerage. Many banks have the physical and computing infrastructure in place to develop with these products and services and an Internet site can serve as an ideal shop front for these services (Jayawardhena and Foley, 2000).

Perhaps, most importantly, Internet banking allows banks to delegate tasks to the customer. Many of the traditional tasks performed by bank counter clerks, and in some cases the domain of line managers, can now be transferred to the customer. For instance, if a customer carries out a transfer of funds between their Internet bank accounts, or pays a bill, they are carrying out a task that would have been undertaken by a bank employee on the customer's instructions in a terrestrial bank. Moving on to higher order transactions, for example, the application for a loan facility. In a traditional setting this would involve a credit officer assessing the credit risks of approving the facility. Now this process can be performed by a customer through their Internet banking services account. Cahoot, a new Internet bank entrant, allows customers to apply for such facilities online, and a decision is given within a matter of minutes. Such transactions save time and money, for both the customer and the bank, the likelihood of committing errors that are the fault of the bank are minimised. Customer's benefit through piece of mind from knowing that transaction details are exactly what they wanted.

2.5 Chapter Summary

This chapter has enhanced the understanding of the banking sector by exploring the characteristics of the sector. This process uncovered threats and opportunities presented by the changing environment in which the banks operate. It was seen that many of the traditional reasons for financial intermediation are being eroded by the combined forces of technology, information, changing markets and regulation. In particular, this analysis highlighted the changing nature of technology in banking and the increased importance and the impact that new Internet bank entrants are having on the banking sector. In continuation of this discussion, the rationale for launching Internet banking services was also discussed.

The emphasis in this chapter is on banks and the banking sector. This descriptive analysis reveals that consumers are faced with an ever increasing choice of banks and banking service products. The combined impact of these developments is likely influence consumer purchase intentions in Internet banking services. Therefore the next chapter shifts the focus from the banks on to the consumer by examining the purchase process, so that a foundation to the understanding of consumer purchase intentions can be laid.

Chapter 3.

The Purchase Process

3.1 Introduction

The last chapter provided the institutional background of the banking sector, which is an important aspect of this thesis. It examined the advent of new Internet bank entrants and technology, and their impact on the banking sector. This chapter shifts the focus from banks to consumers. The objective of this chapter is to lay the foundation for understanding the factors influencing consumer purchase intentions in the Internet banking services of both traditional banks and new entrants.

The chapter is divided into six sections. The first section examines consumer behaviour. Models of consumer behaviour are examined and it is argued that the cognitive consumer model is a useful framework for further analysis. The second section investigates factors influencing purchase intentions. Within these two sections, the literature review takes an inclusive approach by taking into account consumer purchases of both physical products and service products. Subsequent sections narrow the focus on to service products in particular. In the third section, the service product defined and distinguishing characteristics of service products that set them apart from physical products are illustrated. The fourth and fifth sections revisit the cognitive consumer framework in relation to consumer purchase of service products. This discussion also highlights how some aspects of traditional service products are transformed in Internet banking services. Accordingly in the fourth section, attention focused upon consumer behaviour related to the collection and processing of information, and the comparison of the available alternatives. The fifth section extends the discussion by examining the purchase and evaluation of a service product. Such an extended discussion may not be justified in the evaluation of purchase intentions of traditional service products. However the unique aspects of Internet banking services makes this part of the discussion relevant. There are two aspects that deserve attention in this context; (1) the dimensions of a service encounter, which can be used to characterise service encounters, (2) service quality, a construct representing a consumer's overall impression of the relative superiority or the inferiority of the organisations and its services. The penultimate section presents a critique on the interaction of service quality, consumer satisfaction and purchase intentions. This is important, since there are competing arguments and it is important to appreciate the debate and moreover to highlight the stance taken in this thesis.

The chapter concludes with a summary of the key issues.

3.2 Consumer Behaviour

A useful starting point to the analysis of consumer purchase intentions is to examine consumer behaviour. This is an important area, since there are many models of buyer behaviour that have been proposed over the years and an appreciation of these are crucial to the understanding of consumer purchase intentions. Additionally, it is imperative that a suitable model is adopted to examine the research question. It is also equally important to appreciate at this stage in the literature review, no distinction is made between physical products and service products. It prudent to take an inclusive approach initially, for it will enhance the understanding of concepts.

Consumer buying behaviour for goods and services has been extensively examined during the last few decades. Most models of buyer behaviour provide normative rather than strictly quantitative explanations of buyer behaviour and there can be no guarantee that the assumptions on which models were originally based continue to be valid in the context of new environments: for example, Internet banking services. The earliest models of buyer behaviour focused on explaining the decision processes involved in purchases of goods in particular. The following review identifies several explanations of consumer buying behaviour. The different strands of research are allied very closely to the root disciplines of economics, psychology and anthropology.

This generic review will assist in the understanding concepts of buyer behaviour, the purchase process and act as a precursor to examining factors influencing consumer purchase intentions. The distinction between physical goods and service goods is discussed in the fourth section of this chapter.

3.2.1 Models of Buyer Behaviour

Simple models of buyer behaviour usually see an underlying need triggering a search for need-satisfying solutions (Palmer, 1998). When possible solutions have been identified, these are evaluated according to certain criteria. The final purchase decision is a consequence of the interaction between the final decision maker and a range of influences. Finally, after experiencing consumption, consumers develop feelings about their purchases, which influence future decisions.

One widely used framework, which has been applied to consumer purchase decisions, is that developed by Howard and Sheth (1969). This generic model, used to explain purchase decisions, for both goods and services, relied on a framework incorporating: inputs, behavioural determinants, perceptual reaction, processing determinants, inhibitors and outputs. They postulated that, buyers will pass through a cognitive, affective and behavioural stage when there is a high degree of involvement with a product category, which is perceived to have a high degree of differentiation of products within it. During this time, similar models have been put forward by Nicosia (1966), Engel, Kollat and Blackwell (1968), and more recently by Engel, *et al.*, (1991). McKechnie (1992), notes that all these models revolve around the decision making process summarised by Strong's (1925) AIDA (awareness, interest, desire and action) model.

An alternative conceptual framework was put forward by Baker (1983). This model of buyer behaviour was in tum based on Kotler's (1972) framework of four major motivation models (namely Marshallian, Pavlovian, Freudian and Veblenian), which comprise of four different disciplinary explanations of choice behaviour together with six key concepts that were considered to be most helpful in understanding influences which affect choice (namely selective perception; hierarchy of needs; hierarchy of effects; post-purchase dissonance; buy tasks and buy phases; and characteristics of goods). While this model still had the underlying notion of the decision to buy as the outcome of a discrete and sequential process, it endeavours to synthesise key variables in order to provide a useful framework for marketers and academics alike to structure their thoughts and actions around a particular problem so that successful strategies can be developed. Baker (1983) recognised, however, that it would be unrealistic to expect any model to encapsulate completely the complexity and dynamic nature of the buying process, and stated that an additional variable was required to act as a catalyst for the model to work: the specialised knowledge and experience of personal familiar to the specific product market interface being studied.

The importance of the buyer-seller interface has been recognised by the IMP (Industrial/International Marketing and Purchasing) group of researchers. They presented a model that departed from the common approaches. The model conceptualises industrial marketing and purchasing as an interactive process, which takes place within the context of long-term relationships between buyers and sellers. This relationship is central to the buying process. Hence buying and selling, need to be studied in tandem rather than simply look at each aspect separately. The model is built on four factors: (1) both buyer and seller being active participants in the market; (2) the buyer-seller relationship being frequently long term, close and involving a complex pattern of interaction between and within each company; (3) links between parties often becoming institutionalised into a set of roles that each party expects the other to perform; and (4) close relationships often being considered in the context of continuous raw materials or component supply. Essentially, it considers the role of marketing to be the establishment, development and maintenance of relationships between buyer and seller companies (Hokansson, 1982). The constituent components of the model are the interactive process, the participants involved, the environment within which the interaction takes place and the atmosphere affecting or affected by the interaction. As a conceptual framework it enables deeper insights not only into the components of the organisational buying decision-making process, but also ways in which these components interact with each other.

Evaluation methods take a different approach. The multiple-attribute-choice matrix holds that consumers refer to a number of component attributes of a product to evaluate the overall suitability of a product (Palmer, 1998). The evaluation can be carried out in three ways. The simplest method is to evaluate without any weighting of each attribute. The second method involves a linear compensatory approach, whereby consumers create weighted scores for each supplier. The importance of each attribute is multiplied by the score for each attribute. Finally, the third approach to evaluation is called a lexicographic approach (Palmer, 1998). This involves buyers starting their evaluation by looking at the most important attribute and ruling out those suppliers which do not meet a minimum standard. Evaluation is then based on the second most important attribute, with suppliers being eliminated who do not meet their standard. This process is continued until only one option is left.

There is no single explanation of consumer behaviour and intuitively elements from all these approaches can be recognised in one's own behaviour (Gabbot and Hogg, 1998). It is not possible to pursue each of these approaches in this thesis, however and it is necessary to adopt a structured approach to the analysis of consumer purchase intentions. Such an approach tends to favour the information based perspective - a cognitive consumer perspective. That is not to say that the others are either invalid or inappropriate to the consideration of services, rather, the information processing approach offers a clear framework to structure the discussion. The choice can also be defended on the basis that it is the dominant paradigm (Palmer, 1998), and it is both accessible and well documented (Gabbot and Hogg, 1998).

3.2.2 The Cognitive Consumer

The concept of the cognitive consumer is based upon a fundamental agreement that the actual purchase process is a sequence of tasks characterised as a problem solving exercise (Gabbot and Hogg, 1998). This agreement does not in any way dismiss the acknowledgement that there are a number of influences on consumer behaviour. The purchase problem is formulated as a desired outcome where a consumer must evaluate the components of the environment in order to construct a solution or course of action to achieve the desired outcome and comprises of a number of distinct stages. It is appreciated that this model, like most models is a simplification of reality, nevertheless it is a powerful method of analysing individual and group behaviour (Engel, *et al.*, 1991).

The first stage is the increasing recognition of a need or want that can be fulfilled through some form of consumption (Statt, 1997). An important distinction here is in the recognition that not all needs will be translated in to goal-oriented behaviour (Gabbot and Hogg, 1998), i.e. an individual may recognise a desire to apply for a particular credit card, but not be motivated to do anything about it. Alternatively one may have a desire to start a pension fund, but not be able to achieve it.

The second stage in this model comprises of two components. It commences with an information search phase (where consumers collect information about alternatives from memory or advertisements, other people, trade magazines or company databases, and last but not least the Internet), followed by an evaluation stage where consumers engage in some sort of evaluation process to arrive at a ranking of the identified product alternatives (Gabbot and Hogg, 1998). During this stage communication is especially important (Rust, *et al.*, 1996). Providers of goods and services will be concerned to communicate product attributes or benefits via advertising or direct means which closely match those needs sought by the consumer. Similarly the consumer is assumed to be using a wide range of sources and types of information. This process of evaluation continues right up to the purchase of the product. The process may be disturbed if information is presented such as a reduced price, special offer or the unexpected attributes of a previously dismissed alternative (Inman and McAlister, 1993).

The pre-purchase stage of consumption concludes with the purchase itself. According to some authors (Gabbot and Hogg, 1998), the inclusion of this behavioural activity as a separate stage has come under some criticism since it is difficult to extract it from the previous stage. Nevertheless it focuses attention of the physical acquisition of the products, the payment method, delivery, the shop format and role of staff and other people. Additionally it also focuses on the time taken and the human interaction that takes place. While this may be a minor stage with regard to goods, due to the very nature of services, as will be seen later, this is a major consumption stage.

Finally, the post-purchase assessment takes place, once the purchase is complete. The positive evaluation of the purchase, i.e. meeting or exceeding expectations, would lead to satisfaction, a negative evaluation would lead to dissatisfaction. This stage is becoming increasingly important (Gabbot and Hogg, 1998), for a satisfied customer will engage in repeat purchases, and a dissatisfied customer will have a highly detrimental effect on purchases. This is because a dissatisfied customer is often considered to be credible (Parasuraman, *et al.*, 1985 and 1988) and is likely to tell ten or more times the number of people than will a satisfied customer.

This simplified purchase process allows complex behaviour to be broken down into meaningful 'chunks' and allows for analysis (Engel, *et al.*, 1991). Such an analysis also assumes that individuals are fully aware of their actions and act in a rational manner. The decision format identified above is predicted on the assumption that products are inherently similar in terms of the decision process. In reality products are not similar and make different demands upon the consumer. They may be intrinsically different, presented in different environments and evaluated on different bases. Therefore, in any consideration of purchase intentions of consumers, it is necessary to examine the independent factors affecting a purchase, the context in which the consumption takes place and the nature of the product.

3.3 Factors affecting Purchase Intentions of Consumers

The purpose of this review is to understand the factors that may influence consumer purchase intentions in physical products and service products. It will provide theoretical support for the derivation of hypotheses in the fifth chapter. Interdependent forces between price, demand, risk and perceived value determine consumers' purchases of a physical good or service good, the following sections examine these factors in detail.

3.3.1 Price, Demand and Purchase Intentions

Consumer preferences for one product over another are usually represented in neoclassical theory by *indifference curves* (Parkin and King, 1995). An indifference curve shows all combinations of the two products to which the consumer is indifferent. The *marginal rate of substitution* is the rate at which a person will give up one good in order to get another good and at the same time remain indifferent. The marginal rate of substitution is measured as the slope of an indifference curve at a point. It then follows that if the gradient of the indifference curve is steep, the marginal rate of substitution is high, and i.e. the person is willing to give up a large quantity of one good in exchange for a small quantity of the other.

A consumer is assumed to adjust purchases of the two goods until their relative prices (represented by a budget line) are equal to their marginal rate of substitution. Such an equilibrium can be disturbed by a shift in tastes for the two goods, which shifts indifference curves, or by a shift in the budget line as a result of changed incomes and/or prices. If the price of a *normal good* (whose demand increases when income rises) falls, the quantity demanded by the consumer will increase, as indicated by its *price elasticity of demand,* i.e. the responsiveness of the quantity demanded by a good to a change in its price. If the good has perfectly inelastic demand, there will be no change in quantity demanded when price changes. Elastic demand means substantial increases in demand after price reduction. If the consumer's income increases, the demand for normal goods will increase. Inferior goods are goods whose consumption decreases as income increases.

In addition, a change in the price of one product may affect the quantity demanded of another. Income changes can affect both products and price changes themselves influence incomes. The overall demand for any product depends not only on its own price and the level of consumer income, but also on the prices of all other goods and on other variables (such as, brands, advertising, etc.) which may affect consumer preferences (Moschandreas, 1994). The relationship between a product's own price and the quantity demanded is of particular significance both for producers of the product and for consumers. This is because, price may have a strong influence on consumption patterns, and this is given by the *demand curve.* The demand curve is normally negatively sloped.

However, Moschandreas (1994) describes three factors that may influence consumer purchase behaviour besides price and income. *The bandwagon* effect postulates that consumers may buy more of a good when they observe others are buying it. This is because they wish to conform, be fashionable, friendly or simply follow the lead of others, assuming they have collected more information. Whatever the reason may be, consumer purchasing plans are influenced by what others do, which means that they buy more of a good as its market demand expands. The *snob effect* is the opposite of a bandwagon effect, i.e. the desire of the consumer to appear to be different. The snob's pursuit of exclusivity implies that as the total demand for a product increases some snobs leave the market. The *Veblen effect,* or conspicuous consumption, is when consumers buy something just because it is expensive. Purchasing plans are related to a desire to show off so that as the price increases, more is demanded.

The market for a product (physical or service) may include consumers belonging to all three of these groups, the net result will depend on the strength of each group of consumers. For example, if the Veblen effect is stronger than the sum of the bandwagon and snob effects, then the overall market demand can have a positive slope. Apart from these interdependencies, Moschandreas (1994) contends that individuals may behave irrationally, for instance buying on impulse or as a result of whims, or that demand is speculative, in which case demand depends on individual expectations regarding price movements. A price increase may be interpreted as a signal for further price increases in which case speculative demand may increase.

3.3.2 Purchase Intentions and Risk

The analysis so far has an underlying assumption, 1.e. that consumers have full knowledge of all attributes (for example price, supply sources, etc.) of the goods and services that they may be interested in purchasing. In reality, choices (purchases) have to be made among alternatives whose outcomes are dependent on the occurrence of some event which is not known in advance. In such situations where there is an absence of perfect knowledge, consumer choices are said to be based on contingent events (Moschandreas, 1994). If probabilities can be attached to these contingencies, choices are said to be made under conditions of risk. Consumer preferences in situations involving risk are dependant upon, (1) preferences over the outcomes involved, (2) the probability with which outcomes may occur and (3) consumer attitudes with regard to risk. These factors can be combined to construct a von Neumann-Morgestem utility function provided that preferences over the risk-less outcomes are known and that such preferences fulfil the independence assumption (Moschandreas, 1994). The shape of the utility function varies with risk attitudes. Thus, for a risk-averse person, utility increases with wealth at an increasing rate while for risk lovers it increases at a decreasing rate and for risk neutral individuals it increases at a constant rate.

Buyers may be particularly uncertain about the quality of a service or good, and in practise supply conditions and prices themselves may be perceived to indicate quality. This possibility is further considered below in section (3.3.4) under price promotions and discounts.

3.3.3 Perceived Value and Consumer Perceptions

Price and quality have been identified as the two basic components of value in the pricing research literature (Grewal *et al.*, 1998a and 1998b). Dodds and Monroe (1985) proposed a basic model, in which perceived value is positively influenced by perceived quality and negatively influenced by perceived monetary sacrifice. The perceived value of marketers' offerings can be further promoted by comparative price advertising, in which the marketers emphasise that their selling price is a bargain relative to a higher reference price, i.e. the marketer's regular or former price, a manufacturer's suggested retail price, or a competitor's price (Biswas *et al.*, 1993).

Monroe (1979) defined the perceived value as involving a trade-off between perceived quality and perceived sacrifice which may be reinterpreted in the economist's indifference curve/budget line paradigm. Monroe and Chapman (1987) further separated the overall perceived value of an offering into two elements: perceived acquisition value and perceived transaction value. Perceived acquisition value is conceptualised as the buyers' perceived trade-off of the gains or benefits received from acquiring the product or service relative to the sacrifice incurred when making the acquisition. Perceived transaction value is conceptualised as the evaluation of psychological satisfaction or pleasure that would be obtained from taking advantage of the financial terms of the price deal. Grewal *et al.*, (1998b) then developed a model of the effects of comparative price advertising on consumers' perceptions of value. Their proposed model has three exogenous constructs: buyers' perceptions of product quality, advertised selling price and advertised reference price, and five endogenous constructs: internal reference price, perceived acquisition value, perceived transaction value, willingness to buy and search intention.

Subsequently, Grewal *et al.*, (1998b) in their perceived transaction value construct, found that consumers' perceived transaction value is influenced by the advertised selling price. By definition, perceived transaction value involves consumers' assessment of psychological satisfaction or pleasure from taking advantage of a price deal. Furthermore, perceived transaction value has its significance in influencing consumers' final buying decisions. Across the four tests of their model (Grewal *et al.*, 1998b), the relationships between perceived transaction value and perceived acquisition value were found to be significant. Moreover, perceived acquisition value was closely related to willingness to buy. Consequently, an increase in consumers' perceived transaction value can positively affect consumers' perceived acquisition value, which then will in turn enhance their willingness to buy a product.

In Grewal *et al.,' s* (1998a) model, another construct, internal reference price, was also found to be influenced by the advertised selling price. Internal reference price has been defined in various ways, including aspiration, market, and historical prices; adaptation-level, lowest and highest market prices; fair price; normal price perceptions; average market price, expected future price and lowest market price (Biswas *et al.,* 1993; Winer, 1986; Kamen and Toman, 1970). However, all empirical studies of reference prices using scanner data have assumed that past prices are important components of the reference price formation process (Kalyanaram and Winer, 1995). Monroe (1979) argues that internal reference price, existing as a point or a range, is stored in consumers' memory and helps them to evaluate the acceptability or fairness of actual prices. Accordingly, observed prices, i.e. advertised selling price and advertised reference price, are compared against the internal reference price. And, buyers' internal reference price adapts to the anchors (i.e. advertised selling price and advertised reference price) presented in advertisements (Grewal *et al.,* 1998b).

It can be deduced from the above that value is an evaluation that balances what consumers receive in an exchange versus what they give up (Dodds, *et al.,* 1991; Zeithaml, 1988). Thus, essential components of value perceptions include, advertised price, price promotions (or the selling price associated with the price offer) and the perceptions of quality of the brand.

3.3.4 Price Promotions

Previous studies have shown that retail price promotions change consumers' purchase decisions and that retailers use price promotions more frequently to boost store sales (Inman and McAlister, 1993) on the assumption that lower prices will increase demand.

However, most research suggests that price promotions may have a negative effect (over and above inventory effects) on subsequent sales for past users of the brand. (Research that predicts a positive effect generally focuses on nonusers.) This research stems from three very different literatures: (1) reference price theory, (2) self-perception theory, and (3) the economics of information. However, the main implications of the three theories are similar.

Winer's (1986) explanation for a decline in market share after a promotion is withdrawn is based on *reference price* theory. He suggests that consumers set reference prices for brands and that those prices are updated continually as a function of advertising, coupons, and other factors. When coupons or other price discounts such as in-store price promotions are offered, consumers react positively by purchasing the discounted brand. Retraction of the deal, however, disappoints consumers and they may wait until the product is on discount again to repurchase.

Similarly, Thaler (1985) proposes that consumers extract utility from the degree of perceived 'fairness' of a given transaction or sale apart from the utility they derive from acquiring the good. This 'transaction utility' depends on the price the individual pays in comparison with some reference price. The most important factor determining this reference price is 'fairness.' If frequent in-store price promotions change the perceived fair price of the brand, subsequent sales of the brand will be lower. Doob *et al.*, (1969) similarly suggest that consumers tend to think of the product in terms of the price they initially paid for it; if the price increases in subsequent weeks, these consumers see the product as overpriced and are not inclined to buy it at the higher price.

Self-perception theory (Dodson, *et al.*, 1978) suggests that individuals who make a purchase when an incentive is offered are uncertain whether their behaviour is attributable to liking the product (i.e., an internal motivation for purchase) or to a desire to take advantage of the incentive (an external motivation for purchase). When the incentive is retracted, the likelihood of repurchase is reduced because an important reason for purchasing the product is removed. When a purchase is made without an incentive, the only apparent motivation for the purchase is internal and, hence, the inferred positive attitude toward the brand results in increased likelihood of repurchase.

Finally, the economics literature (Akerlof, 1970; Nagle, 1982) provides what is perhaps the most straightforward explanation for why in-store price promotions might decrease subsequent market share for a brand. These researchers suggest that in many marketplace exchanges there is asymmetric information about the quality of the product being sold. For experience goods (Nelson, 1970), whose attributes are determined through use, the seller knows the quality of the goods but the buyers experience great uncertainty. If the buyers cannot identify a seller's product quality, they must rely on some signal to influence their purchase decision. Because price is almost always available and is easy to understand, it is often used as a cue for predicting brand quality. Therefore high price is a market signal for high quality (Huber and McCann, 1982; Spence, 1974); in other words, 'you get what you pay for'. Consequently, a buyer who desires high quality in a market with asymmetric information will be influenced by a price-quality relationship and may assume that a frequently promoted brand is of lower quality. Hence, even if promotions work as an incentive for the promotional period, they may signal that the brand is inferior and the consumer may switch after the promotional period ends.

Though these three theories focus on different aspects of the theoretical process, each can be used to describe the same phenomena. *Reference price* theory stresses that consumers continually update their notion of fair price on the basis of promotional activity, but does not describe the process of how deviations from fair price affect purchase behaviour or how the notions of price affect perceived product quality. In contrast, both *self-perception* theory and *economics of information* theory only implicitly involve the notion that consumers evaluate the fair price of a brand. Rather, *self-perception theory* focuses on how promotions affect attributions for behaviour and hence on future behaviour, and *economics of information theory* focuses on the impact of price promotions on quality inferences. Hence the three theories can be integrated to suggest that though promotions might cause momentary increases in market share, they also may cause longer term decreases in share after they are withdrawn.

3.3.5 Brands and Purchase Intentions

The use of brands is becoming increasingly important as a means of limiting the search activities of potential buyers and of reducing the cost of searching. Rather than considering all possible options, a brand encourages buyers to not consider other products, which do not come with the statement of values, which the brand stands for. Brands are frequently used by companies to provide evidence of consistent standards and are particularly important where a company has not had an opportunity to develop an ongoing relationship with its consumers (Palmer, 1998). Branding has been found to simplify the decision making process by providing a sense of security and consistency which may be absent outside a relationship with a supplier (Barwise and Robertson, 1992).

Brand name is a commonly used extrinsic cue to infer and/or maintain quality perceptions and can represent an aggregate of information about a product (Richardson, *et al.,* 1994). Della, *et* al.,(1981) proposed that a strong brand name helps to control or stabilize the quality perceptions of a branded product even when its price is discounted. In a more recent study, Dodds *et al.,* (1991) also found empirical support for the positive effect of brand name on quality perceptions.

Brand name is also expected to positively influence buyers' internal reference prices. Monroe, *et al.,* (1991) noted that people form an internal reference price scale based on past experience with stimuli. One component of past experience would be recognition of a brand name. Thus, even when consumers have not had direct experience with a particular product or service, exposure to the brand name gives them a certain degree of familiarity. Research supports a positive relationship between brand name and internal reference price (for example, Bearden, *et al.*, 1984). Additionally, brand names may also affect buyers by influencing their internal reference prices through their perceptions of brand quality. That is, past studies have investigated the relationships between actual price and quality (such as, meta-analysis by Rao and Monroe 1989). Thus, brand names are likely to have both a direct and an indirect affect on buyers' internal reference price.

Assimilation-contrast theory explains how an internal reference price might change. Reference prices depend on both product experiences and information in the environment (Zeithaml and Graham, 1993). Consumers form judgements not on the information itself that is given to them, but on their interpretation of that information. As Olson and Jacoby (1977) note, a stimulus is first perceived, then interpreted before it has any effects on judgements and behaviour. Latitudes of acceptance thus exist for price evaluations. A price that is within the latitude of acceptance is assimilated and believable, while a price outside the latitude of acceptance is contrasted, and not credible.

3.3.6 The Impact of Advertising

Advertising can create an awareness of the brand and increase the probability that the brand is included in the consumer's evoked set. It can contribute to brand associations which, when stored in accessible memory, translate into

> *"non-conscious but reliable behavioural predispositions"*
>
> (Krishnan and Chakravarti 1993, p. 214).

Advertising can affect the perceived quality of a brand, and it can influence usage experience.

Stigler (1961) found that advertising which provides information about objective attributes such as price and physical traits will influence brand associations. Both Light (1990) and Nelson (1974) reported a correlation between advertising spending and perceived quality. However they did not find a correlation between promotional weight and the perception of quality. Most of the information content of such advertising, Nelson suggested, is carried in the brand name. In a series of experiments, Kirmani and Wright (1989) found that the perceived expense of a brand's advertising campaign can influence consumers' expectations of product quality.

Advertising can make positive brand evaluations and attitudes readily accessible in memory (Farquhar, 1989). This is crucial to the development of brand equity because, as Herr and Fazio (1992) noted, favourable brand attitudes will only guide perceptions and behaviour if those attitudes can be instantly evoked.

Advertising also influences behavioural manifestations of brand equity. There is a long history of research on the relationship between advertising and sales. On average, market leaders spend 20 percent more of their budgets on advertising than do their nearest competitors (Kim, 1990). Prentice (as cited in Ryan, 1991) compared the effects of consumer franchise building activities (primarily image advertising) versus non-consumer franchise building activities (such as price promotions) on market performance. He found that consumer franchise building (CFB) activities had an impact on sales for about four years, while the effect of non-CFB tactics lasted for one year or less. Johnson (1984a) looked at the relationship between advertising spending and brand loyalty. For those brands that suffered a decline in brand loyalty over time, one of the major contributing factors was a lack of advertising support.

3.3.7 The Moderating Effects of Consumer Knowledge and Prior Ownership

Monroe, *et al.*, (1991, p. 13) noted that,

> *"research on the issue of reference prices must consider the extent that individuals have different degrees of familiarity with the product category and with the different prices charged for various product alternatives ".*

Similarly, Zeithaml (1988) argued that the price-quality relationship is influenced by consumer price awareness and the ability to detect quality variation in a product class. Consumers who are more knowledgeable about product and price information may make different decisions than consumers who are less knowledgeable. Specifically, those who are knowledgeable should be less willing to pay prices that do not reflect the quality of the product compared to those who lack knowledge (Rao and Sieben, 1992). Equally, (Grewal, *et al.*, 1998a) demonstrated that, (1) high-knowledge consumers use brand names to a greater degree to assess perceived quality than do low knowledge consumers; (2) compared to the low-knowledge group, high-knowledge group used less information to make judgements, in other words, the low-knowledge group was swayed by all the information provided.

Product knowledge moderate the effects of price on consumers' perceptions of quality - price had a greater effect for the low knowledge group as compared to the high knowledge group (Rao and Monroe, 1989). Similarly, Biswas and Blair (1991) found that reference price advertisements for an unfamiliar brand affected internal reference price to a greater extent than they did for a familiar brand. A study by Rao and Sieben (1992) discovered that the upper and lower limits of the acceptable price range increased, then levelled off as knowledge increased. They also found that the extent to which intrinsic [for example, product features versus extrinsic (such as, price, brand name)] information was used to evaluate quality varied according to the subjects' knowledge levels. Finally, Laroche, *et al.*, (1996) discovered that familiarity with a brand creates greater confidence, which affects intention to buy the same brand. All these studies, taken together, suggest that consumer product knowledge may moderate the effects of price and other cues (such as brand name) on consumers' internal reference prices, product evaluations and purchase intentions.

This chapter has so far examined models of consumer behaviour and factors influencing consumer purchase intentions of goods. Up to now there have been no distinction between physical goods and service goods, since the objective was to enhance the understanding of principles of consumer behaviour and factors influencing purchase intentions. It is customary to examine concepts in a broad sense and then to narrow the perspective once the first principles are grasped. Hence this is an opportune stage to focus the attention on to service products. The process begins by examining the nature of services and thereafter by appreciating the distinguishing characteristics of services.

3.4 The Nature of Service Products

In order to understand the service product, there is a need to examine what constitutes a product.

A central component in the exchange process, other than the parties involved, is some conceptualisation of a product. This has been defined as,

> *"any bundle of combination of qualities, processes and capabilities (goods, services and ideas) that buyer believes will deliver satisfactions "*
>
> (Enis and Reoring, 1981, p. 17)

It is clear from this definition that product is multi dimensional and dependant upon how the buyer responds to different facets of the offering. Levitt (1976) describes a product as a complex cluster of value satisfactions and identified five levels of product in terms of consumer perceptions. The central part of the product is the *core benefit,* defined as the fundamental benefit or service that the customer is buying. Around the core are four additional product levels: the *generic product, the expected product, the augmented product* and the *potential product.* Each of these describes a different dimension of the product with the capacity to differentiate the offering on the basis of consumer response. Therefore, in reality consumers are dealing with a relationship between business and themselves through the medium of product. In this research only one product class is under consideration - services in general and Internet banking services in particular. It is now appropriate to examine services in detail.

3.4.1 The Service Product - A Definition

Considerable effort has been expended by academics over the last 20 years in establishing that services are different to other products (Berry, 1980; Shostack, 1977; Bateston, 1995) and that these differences present special challenges to the service consumer. However, there is no single universally accepted definition of services (Gabbot and Hogg, 1998; Palmer, 1998). Gronroos (1990) lists a selection of eleven definitions of the term dating from 1960 before arriving - reluctantly - at a definition which he describes as a 'blend' of those suggested by Lehtinen and Lehtinen (1991), Kotler and Bloom (1984) and Gummesson (1987):

> *"A service is an activity or series of activities of more or less intangible nature that normally, but not necessarily, take place in interactions between the customer and the service employee and/ or physical resources or goods and/or systems of the service provider, which are provided as solutions to customer problems"*
>
> Gronroos (1990), p. 27

The complexity and rather convoluted nature of this definition highlights the problems in concisely defining services. Gummesson (1987), referring to an unidentified source, suggests an alternative definition that is more of a criticism of attempts to find and acceptable definition:

> *"Services are something which can be bought and sold but which you cannot drop on your foot"*
>
> Gummesson (1987), p. 22.

Even amongst academics there is no consensuses as to whether the differences between goods and service are significant enough to justify the distinction (Gabbot and Hogg, 1998). As stated above, Levitt (1976) states that there are no such things as service industries, only industries where the service components are greater or less than those in other industries. Shostack (1977) argues that there are very few 'pure' goods or services and describes a product continuum from tangible dominant goods to intangible dominant services. Kotler (1991) provides structure to the continuum by identifying four distinct categories of offering: purely tangible goods, tangible goods with accompanying services, a major service with accompanying goods and services, and pure services. The distinction therefore appears to be, between products where the core of what is being offered is a service and products that are dominantly physical goods (Gabbot and Hogg, 1998). Similarly, Wilson (1972) suggests that the classification should rest on the services relationship to tangible goods. Wilson (1972) goes on to describe three levels of services; (1) services that make available a tangible, for example a supermarket selling groceries; (2) services providing added value to a tangible, for example dry cleaning, and (3) services that provide pure intangibles, for example banking. Although it is possible to argue with some of the examples that Wilson gives within each of his service categories, it is a useful way of considering the service end of the goods-service continuum. The distinction is between products where the core of what is being sold is a service and products where the core is a physical goods with service used as a reinforcement for competitive advantage (Gabbot and Hogg, 1998).

3.4.2 Distinguishing Characteristics of Services

The above analysis shows that the conceptualisation of the term 'service product' is difficult to arrive at, however, all products have characteristics, or attributes, that define the nature of the offering. For services these can be defined as *intangibility, inseparability, heterogeneity, perishability* and *the concept of ownership* (Sasser *et al.*, 1978; Lovelock, 1988; Gronroos, 1978).

3.4.2.1 Intangibility

Intangibility is the absence of physical dimensions. It is one of the most important characteristics of service products. Often services are described using tangible nouns but this confuses the fundamental nature of the service which remains intangible (Gabbot and Hogg, 1998). Berry (1980) describes a good as 'an object, a device, a thing', as opposed to a service which is 'a deed, a performance, an effort'. He goes on to argue that even though the performance of most services is supported by tangibles, the essence of what is purchased is a performance. McLuhan (1964) concurs by pointing out that, it is the process of delivering a service which comprises the product. The implication of this argument is that services are low in visibility, cannot be touched, lacks taste or smell; services are high in *experience qualities,* which refer to attributes which can only be assessed after purchase or during consumption (Carman and Uhl, 1973; Sasser *et al.*, 1978). By extension it follows that, as Bateson (1977) asserts, at the pre-purchase stage, services are more difficult for consumers to evaluate than goods, since any evaluation will be low in *search qualities,* which are tangible attributes and can be considered in advance. Furthermore, many professional or specialist services will also be high in *credence qualities,* which are attributes which cannot even be assessed after purchase and consumption (Zeithaml, 1981).

3.4.2.2 *Inseparability*

Inseparability service is a performance, where the production and consumption aspects of a service transaction takes place in real time, with the purchaser cooperating with the provider (Bell, 1981). This is referred to as inseparability (Gronroos, 1978). Thomas (1978) argues that the degree of this involvement between the transacting parties is dependant upon the extent to which the service is 'equipment based' or 'people-based'. Equipment based services (such as telephones, vending machines, etc.) deliver the same service to whoever consumes them. On the other hand, people-based services where the involvement of a human service provider inevitably means that individual consumers will experience variation in the service delivery. The implication of this distinction is that people-based services tend to be less standardised than equipment based services or goods producing activities (Gabbot and Hogg, 1998). Goods are produced, sold and then consumed, whereas services are sold and then produced and consumed simultaneously (Regan, 1963 and Cowell, 1984). The inseparability of the role of the service provider and consumer therefore leads to a lack of standardisation since both the purchaser and the individual delivering the service can alter both the way in which the service is delivered, as well as what is delivered. This process has important implications for the process of evaluation (Gabbot and Hogg, 1998).

3.4.2.3 *Heterogeneity*

This characteristic of services is related to the above characteristic. Services are produced and delivered by individuals and therefore each service encounter will be different by virtue of the participants, the time of performance or the circumstances consumed at the same time (Gabbot and Hogg, 1998). As a consequence each purchaser is likely to receive a different service experience, or in other words the service is heterogeneous. In the case of goods it is possible to standardise the product and guarantee replication of the same product over time, but this is not possible in services. A great deal of the services literature has been concerned with mechanisms for achieving a degree of uniformity in service delivery, for instance, Levitt (1972 and 1976) considers the industrialisation of services, and to an extent the services quality literature, is also based upon some standardisation in delivery (Gabbot and Hogg, 1998). On the other hand, where individuals and interpersonal exchange are involved there may well be attractions in having personalisation and customisation.

3.4.2.4 *Perishability*

The perishability of services describes the real time nature of the product (Gabbot and Hogg, 1998). Unlike most goods manufacturers who can normally hold stocks to meet fluctuating levels of demand, services cannot be stored (Palmer, 1998), i.e. the consumer cannot stockpile services against a possible need to in the future. In certain circumstances the consumer may decide to delay consumption but not to consume more in advance or requirements. For the purchaser of services the time at which they choose to use the service may be crucial to its performance and therefore to the consumer's experience (Gabbot and Hogg, 1998). For example, the experience of carrying out a transaction at a bank in the lunch hour from other times of the day. Kelly, *et al.*, (1990) make the observation that consumption is inextricably linked to the presence of other consumers and their presence can influence the service outcome.

3.4.2.5 *Ownership*

The concept of ownership is a distinguishing feature of services (Judd, 1964; Kotler, 1982). In contrast to the case where with the sale of a good the purchaser generally obtains ownership of it, in the case of a service the purchaser only has temporary access to or use of it. What is owned is the benefit of the service, not the service itself, for example when a bank customer withdraws cash from a bank ATM, the customer has the benefit of the machine but does not own them. The absence of ownership stresses the finite nature of services for purchasers - there is not enduring involvement in the product, only in the benefit (Gabbot and Hogg, 1998).

The second section of this chapter introduced the concept of the cognitive consumer, and it was argued that this represents a robust framework to analyse the consumer purchase process. It was seen that a cognitive consumer makes a purchase decision in a sequence of stages. The preceding section defined services and examined their characteristics. Now it is opportune to examine in detail, the stages that a consumer goes through in a service good purchase process. This exercise is important in highlighting any differences that may exist in comparison to physical good purchases. It will enhance the understanding the principles of consumer purchase intentions m services and by extension consumer purchase intentions in Internet banking services.

3.5 Pre-Purchase Stage of Services

This section examines consumer behaviour that takes place prior to the contact with the service provider, in particular behaviour related to the collection and processing of information, and the comparison of the available alternatives.

3.5.1 Information Acquisition

Faced with any purchase decision, a consumer requires information with which to inform product choice. This is usually referred to as an information acquisition phase which commences with a search of already held information (Gabbot and Hogg, 1998). Bettman (1979) and Jacoby, *et al.*, (1978) characterise this process as a scan of memory for experiences which constitute a body of knowledge about, or an attitude towards, a product or product class. Once the consumer determines that this information is insufficient, for example, by virtue of its volume, age, specificity or accuracy, the consumer embarks on an information acquisition phase. The first step in this process is the identification of desired information from external sources.

The degree and the intensity of the information search is said to be dependant upon a number of factors, such as product category experience, product complexity or the degree of buyer uncertainty and on each of these dimensions, services are likely to prompt significant external search effort (Gabbot and Hogg, 1998). The literature on consumers' external information activity is large and covers many facets; Engel, *et al.*, (1991) concentrates upon classifying the various sources of information, Keller and Staelin, (1987) and Jacoby, (1984) discuss the ability to assimilate information from these sources, and Bucklin, (1996) and Urbany and Weilbaker, (1987) concentrate on the motivational aspects for external search behaviour and the extent to that behaviour. In considering the degree of information search in services it is inadequate to simply analyse the absolute number of sources used, but a more prudent and productive approach is to assess source effectiveness (Gabbot and Hogg, 1998). Murray (1991) argues that this approach is particularly appropriate for services because of the additional information burden on consumers. This burden or information overload as it is referred to, is associated with the sources of information, the nature of information available from each source and the consumers' response to that information.

It will be seen in section (5.3), that there are a number of methods through which a consumer may acquire information in Internet banking services, including, (1) by relying on the contents of the Web site, and (2) through the demonstration component by simulation of account purchase and operation. Accordingly, information acquired through the simulation processes is particularly high in experience qualities and this is an important aspect in Internet banking services.

3.5.1.1 Information Effectivenss

In order to evaluate the information demands of services, Zeithaml

(1981) after Nelson (1974) and Darby and Kami (1973) suggests a framework based on the inherent qualities of products. This framework was first applied in the economic regulation literature and uses three categories of product qualities: *search qualities,* which are attributes a purchaser can determine prior to purchase; *experience qualities,* which are attributes which can only be determined after purchase or during consumption; and *credence qualities,* characteristics which purchasers may find impossible to evaluate even after consumption. All products, it is suggested, can be described in terms of the proportions of the three qualities (Gabbot and Hogg, 1998). Services are characterised as being low in search qualities but high in experience and credence qualities (Zeithaml, 1981; McKechnie, 1992).

According to Gabbot and Hogg (1998), although this conceptualisation of service information demand is useful, a number of shortcomings in this framework have not been adequately addressed to date. Firstly, the terminology is confusing. For instance, the conceptualisation of products possessing 'qualities' as opposed to attributes or characteristics, and in particular the term 'search'. This term does not describe information search behaviour as such, but a characteristic of the information. This means that the information is searchable, or as Nelson (1974) refers to it *"inspectable prior to purchase"* (p. 312). In the case of services, searchable qualities are said to be those which are associated with tangible sources of information. This searchable information may not necessarily describe the nature of the service experience which remains intangible. Secondly, there is a vague distinction between search and experience qualities. For example, Zeithaml (1981) states that purchasers can obtain information about experience qualities in the pre-purchase phase by using the experience of others. This has the effect of making experience 'searchable', since the source is tangible and can be consulted prior to purchase. Finally, the importance of the credence characteristics and the difficulties in evaluating certain services even after consumption means that purchasers are likely to find some way of approximating the missing information.

Searchable Information

Nelson (1974) defined search qualities as attributes that can be determined prior to purchase. They are usually associated with factual information which is common to all consumers and verifiable in advance of purchase. It is important to note that the information do not directly describe the experience which the consumer will actually receive. This in tum is central to the pre-consumption phase problem (Engel, *et al.*, 1991). However there is very little consistent evidence concerning consumer responses to this situation (Gabbot and Hogg, 1998). For example, consumers will seek to simplify their purchase decisions. At the same time, they would also want to minimise the level of risk attached to the outcome by collecting as much information as possible.

Bettman (1975) suggests that a possible explanation of this apparent contradiction the consumer's simultaneous attempt to increase the value of the information held and to reduce its volume. Gabbot and Hogg (1998) argues that one method of achieving this outcome is to use the searchable characteristics of services to estimate experience or credence characteristics. Zeithaml (1988) extends this process further by using an intrinsic/ extrinsic distinction to examine product information. The relationship between searchable and non-searchable attributes can be explored by drawing a distinction between intrinsic information (objective product features) and extrinsic product information (part of the product value but which can be changed or removed without fundamentally altering the product). Extrinsic or subjective attributes are used to approximate missing intrinsic information, for example price to approximate quality or brand to approximate reliability.

There is very little intrinsic information in services. On the other hand, there is likely to be a large amount of extrinsic information that can be used to infer product performance such as the appearance of the service provider, the premises, or quality of equipment (Zeithaml, 1988). This extrinsic information is searchable within the definition provided by Nelson (1974). Therefore, it can be said that services do have searchable qualities. However, these qualities are not intrinsic to the product. For instance, in describing the service which the company can or will provide, this searchable information may be used by the consumer to give indications as to the likely service experience. However, an important consideration associated with the argument in favour of product cues for services, is in their predictive reliability (Gabbot and Hogg, 1998).

Where there are tangible cues, these tend to vary from provider to provider and consumer to consumer (Bettman, 1979). Therefore, by extension the effectiveness of cues is likely to vary between transactions. While product cues in relations to goods are used at the pre-purchase stage and their value assessed in the post purchase stage, where services are concerned, the actual delivery may take place at a different time, with a different provider, with different tangibles or in a different place to the purchase transaction (Keller and Staelin, 1987). Therefore, cues can be used to evaluate a service at the pre-purchase stage may be different from those used in evaluation during delivery or even after delivery has taken place (Gabbot and Hogg, 1998).

The second assumed response, according to Gabbot and Hogg (1998), of consumers to an information shortage is brand or product loyalty. This in effect is one form of routinising purchase behaviour. In the case of services, loyalty can only be placed with the provider of the service rather than the service itself. The only way to build individual loyalty is to experience a series of successful service encounters with the same provider, the aggregation of all these consumers with successful encounters builds the reputation of the service provider. Gabbot and Hogg (1998) asserts that one of the most important aspects of loyalty with respect to services is the ability to form long lasting relationship between the provider and the customer. Additionally, with each service encounter results in an incrementally strengthening service relationship. In the case of goods, such an incremental relationship building process is unlikely to take place, instead the relationship may remain static or even start to decline, depending on the circumstances.

3.5.1.2 Experience Based Information

Many authors (Gabbot and Hogg, 1998; Bettman, 1975 and 1979) regard experience as the most effective source of consumer information in the pre-consumption stage. The heterogeneous nature of services means that it is not possible to guarantee a repeat performance. Nevertheless experience based information is extremely credible and may relate to the service provider, the service class or similar service experiences (Gabbot and Hogg, 1998). The consumer is faced with the problem of having to obtain experimental information at the pre-purchase. The only sources of this type of information are (1) the consumer could have used the service provider before for a similar service experience, (2) a pre-purchase trial, (3) observation of others receiving the service experience, (4) reliance upon the experiences of others (Locander and Hermann, 1979).

A consumer could have used the same service provider for the purchase of other financial service products, and this process may give the consumer the necessary experience to what to expect from the provider.

Pre-purchase trial is not a common option in the case of services although it is possible to try one service episode to establish whether one will purchase a full service or a sequence of episodes. However, unlike a physical product where the trial is exactly equivalent to the actual product, in services there is no guarantee that the rest of the service bears any relation to the experience of the trial (Gabbot and Hogg, 1998). The service is intangible and the participation of any other individual gives no guarantee of a repeated performance hence observation may be unreliable as a source of information. As a consequence, a number of authors suggest that consumers look towards experiences of others (Murray, 1991). This position is supported by Zeithaml (1981), who suggests that the need for experience information of the service prompts reliance upon word of mouth sources as they are perceived to be more credible and less biased. In addition the work of Robertson (1970), Eiglier, *et al.*, (1977) and Urbany and Weilbaker (1987) indicate that word of mouth sources are pivotal in relation to services.

As a consequence it can be concluded that where service is a dominant element of a product, consumers face a number of problems in acquiring and using experience, and that the likely response is an increased reliance on personal sources of information (Gabbot and Hogg, 1998).

As explained above, the availability of demonstrations in Internet banking services provide high experience qualities through pre-purchase trials (by simulation). It is also the case that, subsequent service experiences are near identical to the experience gained through simulation. This is an important distinction of Internet banking services that may be lacking in traditional services. These aspects are examined in detail in chapter five.

3.5.2 Comparison

The process information search leads the consumer to an evoked set of alternatives that are likely to fulfil their expectations which form the basis of comparison and choice (Gabbot and Hogg, 1998). For reasons outlined above, it is difficult to obtain effective pre-purchase information about services. Hence in comparison to goods, services are likely to result in a smaller evoked set. Hence, Zeithaml (1981) argues that consumers tend to be more loyal once they have found an acceptable alternative.

If the consumer has previous experience of a service the evoked set may be as small as one (Johnston and Bonama, 1981). On the other hand, if the experimental information is negative, or the consumer does not have experience on which to base the choice, then the size of the evoked set will be dependant upon the effectiveness of the external information available.

A number of authors, including, Bettman, (1979), Grether and Wilde (1984), Fletcher and Hastings (1983) have presented various models of how consumers choose between available alternatives in different situations. The common denominator of these models is a set of attributes. However, there are problems for consumers in defining attribute sets in relation to services. To begin with there are problems of identifying what the attributes are in the first place. If the first hurdle can be overcome, there are problems in making comparisons on the basis of these attributes.

Both goods and services have attributes or defining characteristics, with the attributes of goods being tangible and those of services are intangible. Tangible attributes can be determined prior to the purchase and are common to all consumers purchasing the product. Intangible attributes, by contrast cannot be determined in advance of purchase, and are not common to all consumers. A consumer's needs are accommodated by their involvement in the service delivery, hence the attribute will be a function of that individual involvement. For example, when a bank customer encashes a cheque, the customer is involved in describing how they want the cash, i.e. the denomination mix, thus modifying the service outcome. In the absence of any tangible indications of that the service will be like, consumers must use other means of comparing services in the pre-purchase phase (Gabbot and Hogg, 1998). Shostack (1977) and Berry (1980) argue that consumers rely upon peripheral tangible cues to predict the likely quality of the service. Levitt (1981a and 1981b) argues, that in instances where the service is intangibly dominant, it is necessary for consumers to establish metaphors for tangibility or cues that help them to 'tangibilise the intangible'. In doing so the consumers may be able to create a credible expectation. Bitner (1992) and Lewis (1991) has pointed to the role of the environment in which the consumption of the service takes place in providing these metaphors. Consumers may use corporate wear, decor, appearance of service providers, and the standard of equipment or furnishing in approximating the missing tangible product information. However, it is difficult for the consumer to identify the cues which will most accurately predict the nature of the service experience.

As identified above, consumers find it difficult to compare service alternatives on the basis of common attributes. A consumer cannot put two services side by side at any one time, hence they cannot be compared simultaneously, but can only be compared in series, not parallel. An additional problem of comparison is the problem of heterogeneity. The absence of truly common attributes implies that services are non-comparable products. When a consumer is confronted with non-comparable product alternatives they will search for the basis of a comparison by moving to more abstract product attributes (Johnson, 1984b), for example necessity, social status or entertainment value. In the case of services, non-comparability is likely to evoke a reverse from of abstraction where services are compared on increasingly material or tangible criteria until there is little left to compare other than the service provider as the ultimate physical embodiment of the service (Gabbot and Hogg, 1998).

Herein lies an important distinction between traditional service goods Internet banking services. As stated earlier, the demonstration component enables consumers to simulate the purchase and operation of Internet banking service accounts. Hence, it is possible to compare service alternatives based on common attributes. Consumers can, if they so wish, put two Internet banking services side by side and therefore can be compared simultaneously.

A further characteristic of service products is that some attributes are bargainable (Gabbot and Hogg, 1998). These bargainable attributes are determined between provider and consumer by negotiation. Brucks and Shurr (1990) define bargaining as a process whereby two or more parties mutually define one or more attribute values for a product. The bargainable nature of some service attributes serves to emphasise the uncertainty of the comparison process. This factor also has implications for the number of alternatives as well as significantly reducing the number of attributes used in the comparison process (Gabbot and Hogg, 1998).

What is being assessed in the case of a service is the perceived benefit from the service, rather than the service itself. The consumer is choosing between their own subjective assessments of the likely service outcome. Comparison is further hampered by the heterogeneity of service provision and the difficulties in identifying or generating attributes upon which to base a choice.

3.6 The Purchase and Evaluation of a Service

The previous section examined consumer behaviour that takes place before contact with the service provider. This section moves the discussion further with the examination of the purchase and evaluation of a service. There are two aspects that merit attention in this context; firstly, the dimensions of a service encounter, which can be used to characterise service encounters, and secondly, the service quality, a construct representing a consumer's overall impression of the relative superiority or the inferiority of the organisations and its services.

This review may not be justified if the research question is examining the characteristics of a traditional banking services that influence consumer purchase intentions. However, as already stated earlier in this chapter and examined in detail in chapter five, the demonstration component of an Internet banking services provides an opportunity for consumers to purchase and experience account operation by simulation. Therefore the various dimensions (that can be experienced through simulation) and service quality of an Internet banking service encounter may influence consumer purchase intentions. Therefore this discussion is pertinent to the thesis.

3.6.1 Dimensions of a Service Encounter

Various dimensions of a service encounter influence consumer behaviour. Seven common dimensions applicable to all service encounters can be identified in the literature: the role of employees (Zeithaml and Bitner, 1996), the service context or the environment (Bitner, 1992; Zeithaml and Bitner, 1996), the physical proximity of participants (Bitner, *et al.*, 1990, Zeithaml and Bitner, 1996), the degree of participation (Zeithaml and Gilly, 1987; Pine and Gilmore, 1998), the temporal duration (Lovelock, 1995), the degree of customisation (Pine, 1993; Gilmore and Pine, 1998), and the degree of cognitive involvement (Price, *et al.*, 1995; Murray, 1991). The effect upon consumers of each of these dimensions is considered in tum.

3.6.1.1 Employees

A characteristic that sets service product from other products is the role of employees or people creating the service experience for the consumer. All types of employees in an organisation, i.e. both front office and back office staff, work in tandem to deliver the consumer's service experience. While in some services, a number of employees are involved in delivering the service, such as, a journey in a plane, in others an individual will undertake all tasks associated with the service, for example, financial advice.

Employees possess three characteristics that directly impact upon the consumer's experience (Czepiel, *et al.*, 1985). Firstly, the expertise of the employee in securing the desired service outcome. Employee expertise may comprise of two aspects. On one hand this may be associated with formal qualifications that the employee has gained, which by definition can be assessed; and increasingly, on the other hand, from the consumers perspective are abilities which are associated with creativity, flexibility, and response to the customer. Secondly, the individual character traits of the employee, including their openness, friendliness, empathy, etc. These individual characteristics are also identified by Hochschild (1983) and Ashforth and Humphrey (1993). The third and final characteristic is the demography of the service provider. Although it is politically incorrect to raise issues such as, gender, age educational level of the employee, all these have a direct effect on the consumer experience.

Bank employees are not normally directly involved m an Internet banking service transaction. However, direct input of bank employees is required at account opening stage, for verification (for example, identity, address, etc.) purposes. The present regulatory framework which has been designed for terrestrial banking requires these manual verification processes. Other instances where bank employees interact with customers may be during the account operations, where a customer may require assistance.

What is important is that unlike in a traditional service encounter, the employee-customer interaction is remote. What is important is the ability of the Internet bank's employees in delivering the desired service outcome. The importance of Individual character traits that are crucial in face-to-face contact is subdued in Internet banking services. This has implications both for consumer purchase intentions and also the evaluation.

3.6.1.2 *Service Context*

The environment in which the service takes place will have a direct impact upon the consumer's perception (Bitner, 1992; Zeithaml and Bitner, 1996). However, despite the importance of the service environment on the consumer, there has been little explicit research conducted by service researchers outside retailing (Gabbot and Hogg, 1998). A possible explanation for this void is the difficulty in isolating the environmental effects from other variables associated with the service encounter.

The environment that consumers receive the intended service in most traditional service products is within the control of the service provider. The environment in which an Internet banking service products are received can be divided into two components: the physical environment and the virtual environment. The physical environment, comprising of the physical location of the computer and associated peripherals are located, is principally controlled by the consumer. It is the consumer that chooses to use a certain computer, at a particular time at a particular location. The virtual environment, on the other hand, is in part controlled by the Internet bank, and aspects such as the quality of the Internet connection (which can be construed to be a part of the virtual environment since it affects the quality of the environment) may be controlled by other parties (for example an Internet service provider, etc.). Hence it could be argued that an important dimension that was previously controlled by the service provider, and therefore used to create an impact on consumer perceptions, has been taken away from the Internet banks to a considerable degree; the Internet banks may have lost a considerable degree of influence over consume purchase intentions.

3.6.1.3 *Proximity*

A service encounter takes place every time a customer interacts with a service organisation (Gabbot and Hogg, 1998), these encounters, however, do not necessarily require the parties to physically meet. Three types of encounters can be identified with regard to the physical proximity of the service provider and the consumer; face-to-face encounters, remote encounters and telephone encounters or remote personal encounters (Zeithaml and Bitner, 1996).

In traditional service delivery, face-to-face encounters are the norm. The human interaction component is central to service quality (Bitner, *et al.*, 1990) and customer satisfaction (Quelch and Ash, 1991; Brown and Swartz, 1989; and Surprenant and Solomon, 1987). In a face-to-face encounter, the parties are involved both consciously and subconsciously in a form of social action (Blumer, 1953). There is a richness in the interpersonal communication between the service provider and the customer. Further, there is a larger degree of participation, and by definition enhances the service product, or in other words the service product is customised.

In a remote encounter, the interaction between the customer and the service organisation is mediated by using some form of technology, such as, an ATM machine or the computer and the Internet in the case of Internet banking services. By definition, these encounters lack the interpersonal aspects highlighted above. The consumer is faced with a number of behavioural boundaries, which are in tum dictated by the level of technology. The consumer has to be self reliant to a large degree in this form of encounter, and must formulate service episodes into behavioural sequences (Gabbot and Hogg, 1998).

In a remote personal encounter, personal interactions take place without the need for physical proximity. Although this form of encounter is not as rich as face-to-face encounter, it does contain some advantages over a remote encounter, including; the ability of the consumer to negotiate, customise, monitor service performance, etc. This form of encounters was primarily conducted over the telephone; however with the development of communication technologies, such as video conferencing, Internet enabled cameras, etc. there has been a significant increase in the quality of communication.

3.6.1.4 *Participation*

The concept of participation is closely related to the above notion of proximity. Customer participation is essential for a service encounter to take place, and it is possible to identify degrees of involvement. Gabbot and Hogg (1989) states that customer participation in the service product creation process can vary from passive to active. This dimension is also closely associated with concepts of customisation.

Technology plays an important role in this dimension and has brought about increasing changes to the role of customer participation with regard to service delivery (Stem, *et al.*, 1993; Bateson, 1985). Traditionally, the customer interacted with the service delivery personnel in a particular service context. Increasingly, customers are being asked to increase their level of participation with the tangible physical aspects of the service environment and reduce the level of interaction with the provider (Gabbot and Hogg, 1998). For example in an Internet banking services transaction procedure the consumer has taken over many tasks performed by a bank clerk in a traditional banking environment. This shift in the level of participation would undoubtedly lead to the nature of the service itself, since many researchers have concluded that one of the most important characteristics of the service product is the role of the customer in its delivery (Parasuraman, *et al.*, 1985; Bitner, *et al.*, 1990; Gronroos, 1978). Indeed, research by Gabbot (1996), Zeithaml and Gilly (1987), Hiltz and Johnson (1990), suggest that participation is as much a consumer variable as it is a product attribute. This group of researchers further suggest that the changing nature of participation may have a number of as yet unidentified benefits and costs for consumers.

3.6.1.5 Duration

Services encounters on one hand can be long and repetitive and both short and discrete on the other. Gabbot and Hogg (1998) allege that there has been a tendency in published research to concentrate upon short and discrete encounters simply because they offer less methodological and conceptual challenges. This is a serious shortcoming in light of the fact that most important consumer services, such as banking services, may be characterised with long encounters with many interactions. In single encounter services, the consumer is experiencing the purchase of a service product along the lines of a purchase of a physical good. Where the service entails a series of interactions, for example, an estate agent, the transaction is completed over a number of service encounters before the exchange is complete. Even a single encounter can be broken down into small individual interactions, and all these different individual interactions offer different and distinctive experiences with a particular space of time (Gabbot and Hogg, 1998). An analysis of all these encounters can assist in the understanding of the consumer experiences and provide a contribution to perceptions of service quality and satisfaction.

Once a consumer purchases an Internet banking service it is likely that there may be many interactions with both long and short encounters (depending on what aspect of the service that the consumer wish to receive). As stated elsewhere, prior to the purchase of an Internet banking services a consumer is able to assess the likely duration of encounters by simulations. Therefore, in formulating the research methodology to test the hypotheses, it is important to address the criticisms highlighted by Gabbot and Hogg (1998), and appreciate the nature of the duration dimension. This aspect is examined in chapter six.

3.6.1.6 Customisation

Unlike physical goods where products are homogeneous, service products are heterogeneous. There are two dimensions are associated with customisation; (1) the degree to which the consumer interacts with the service process, and (2) the degree to which the service is customised. From the consumer's perspective, customisation is a function of how the service personnel are willing or able to respond to the consumer's individual demands (Gabbot and Hogg, 1998).

The fact that most services are customised is presented as a positive aspect of the service product. However, it does present a number of challenges to consumers (Pine, 1993). Firstly, in order to demand a customised service product, the consumer must have a high degree of product knowledge. Secondly, the consumer must be sufficiently competent and confident to intervene in the service delivery process. This may present problems, since in most instances the service provider is an expert, for example, in the health service it is difficult for a patient to influence a doctor. Thirdly, where the service provider advises against customisation, there is a significant risk in initiating customisation. In such instances the provider might refuse outright. Alternatively the consumer might suffer from retribution from the provider for making their task more complex or less standardised (Gabbot and Hogg, 1998). Therefore, while customisation is seen as a consumer benefit, in practise achieving it may present difficulties.

It could be argued, as will be examined in detail in section (4.5), that Internet banking services by their very nature make the service more of a homogeneous product and that it is difficult to customise along criteria used in traditional banking services. The implication is that Internet banks are forced to customise their products by emphasising more on characteristics such as interest rates, charges, etc. These characteristics can be evaluated using objective criteria. Therefore it will argued in chapter six, that the evaluation of characteristics of Internet banking services that influence consumer purchase intentions can become more objective, compared to traditional banking services.

3.6.1.7 Involvement

This construct, the consumer engagement with a product, be it cognitive, emotional or behavioural, is widely used in explaining different degrees of activity in relation to product choice and consumption (Gabbot and Hogg, 1998). The level of engagement will vary with the type of service. Apart from the type of service, factors such as mood, personality and individual resources will also play a part in determining the degree of interest shown by the consumer of a particular service. It was seen above that there can be several levels of proximity in a service encounter, but every service encounter will take place within a service context, and at a certain level of engagement, which then becomes an integral part of the service encounter experience, whether or not it is intended to be so. Although this aspect of a service encounter will have a major impact upon both the consumption and post-consumption experience, Gabbot and Hogg (1998) note that this area of a service encounter has not received much academic attention, other than an article by Bitner (1992).

Consumer involvement in Internet banking services is very high, and this dimension is explored in more detail in section (4.5). As explained earlier many of the tasks performed by bank clerks in traditional banking services have been transferred to the consumer, thereby making the service experience largely a function of how consumers carry out these tasks.

3.6.2 Service Quality

Quality is an important basis which consumers use for differentiating between competing services. Unlike tangible goods, the quality of a service can normally be assessed experimentally, i.e. only once a service has been consumed. As reiterated a number of times in the above discussion, the demonstration component of an Internet banking service along with other components provide an opportunity for the consumer to experience the service through simulation. Therefore, before actually consuming the service, consumers are able to assess service quality dimensions.

The discussion begins by defining the service quality construct.

3.6.2.1 A Definition

Quality is a difficult, elusive and indistinct construct to define in a few words, and it is more so in the services sector. Often mistaken for imprecise adjectives like 'goodness' or 'luxury', or 'shininess', or 'weight' (Crosby, 1979), quality and its requirements are not easily articulated by consumers. A further problem in defining service quality lies in the importance which consumers often attach to the quality of the service provider as distinct from its services; as noted earlier, the two cannot be separated as easily as in the case of goods (Palmer, 1998).

Efforts in defining and measuring quality have come largely from the goods sector (Parasuraman, *et al.,* 1985). At its most basic, Crosby (1979) defines quality as 'conformance to requirements'. This implies that organisations must establish requirements and specifications; once established, the quality goal of the various functions of an organisation is to comply strictly with these specifications (Palmer, 1998). Garvin (1983) measures quality by counting the incidence of 'internal' failures (those observed before a product leaves the factory) and 'external' failures (those incurred in the field after a unit has been installed). Parasuraman, *et al.* (1985) defined service quality as the degree of discrepancy between consumers' normative expectations for the service and their perceptions of the service performance, where 'normative' refers to what consumers believe a service provider should offer. Palmer (1998, p. 34) combines previous definitions of service quality:

> *"Quality can only be defined by consumers and occurs where an organisation supplies goods or services to a specification that satisfies their needs".*

There is a considerable literature that highlights the relationship and confusion between attitudes, satisfaction and service quality. Parasuraman, *et al.*, (1988) and Bolton and Drew (1991a) describe service quality as a form of attitude, related but not equivalent to satisfaction, that results from the comparison of expectations with performance. The most common explanation of the difference between these is that perceived service quality is an attitude, a long run evaluation, whereas satisfaction is a specific measure. However, this differentiation is inconsistent with Woodruff, *et al.*, (1983) and Bitner (1990). As will be seen in a later section, this confusion and apparent lack of agreement has important implications for the measurement of service quality.

There is a substantial literature on service quality (Gronroos, 1982 and 1984; Parasuraman, *et al.*, 1985, 1988, 1991, 1994a and 1994b; Cronin and Taylor, 1992 and 1994; Boulding, *et al.*, 1993; Brown *et al.*, 1993; Rust, *et al.*, 1995). These writers suggest that service quality is different from goods quality, because most services have unique characteristics that set them apart from goods, and service quality is more difficult for the consumer to evaluate. Secondly, service quality perceptions result from a comparison of consumer expectations with actual service performance. And, thirdly, quality evaluations are not made solely on the outcome of a service, they also involve evaluation of the process of service delivery.

3.6.2.2 *Service Quality Characteristics and Dimensions*

Analyses of service quality have highlighted the unique characteristics and dimensions of service quality, these are summarised in Table (3.1, next page). Many have attempted to distinguish between objective measures of quality and measures which are based on more subjective perceptions by consumers. In practise, the line between objective and subjective measures is blurred.

Quality Dimension	Author(s)
Instrumental - physical aspects of the service *Expressive* - intangible or psychological aspects	Swan and Combs (1976)
Technical - relatively quantifiable aspect of *what* a consumer receives as a result of his interaction with the service firm *Functional* - *how* a consumer receives the outcome of a production process. It corresponds to the expressive performance of a service	Gronroos (1984)
Material facilities and *personnel.*	Parasuraman, *et al.,* (1985) [quoted from Sasser, *et al., (1980)*]
Physical quality - the physical aspects of the service *Corporate quality* - the company's image or profile *Interactive quality* - which is a function of the interaction between the contact personnel of the firm and the consumer, as well as between some consumers and other consumers.	Parasuraman, *et al.,* (1985) [quoted from Lehtinen and Lehtinen (1978)]
Tangibles *Reliability* *Responsiveness* *Communication* *Credibility* *Security* *Competence* *Courtesy* *Understanding/knowing the consumer* *Access* (These 10 dimensions were derived during the initial development of the SERVQUAL scale)	Parasuraman, *et al.,* (1985)

Table (3.1): Service quality dimensions

Services can be seen as not objects, but rather performances, with the consequence that precise manufacturing specifications concerning uniform quality can rarely be set (Parasuraman, *et al.*, 1985, Bateston, 1977). It is apparent from the above table that both the consumer and the supplier can measure some dimensions, but others can only be assessed by the consumer. Service quality involves not only the outcome, but also the manner and the process in which it is delivered. The intangible process characteristics which define services, such as reliability, personal care, attentiveness of staff, friendliness, etc. can only be verified once a service has been purchased and consumed. Because services are heterogeneous, their performance varies from one provider to another, from one consumer to another, and from day to day (Parasuraman, *et al.*, 1985). The opportunity for pre-delivery inspection and rejection which is open to the goods manufacturer is not normally possible with services: service must normally be produced in the presence of the consumer without the possibility of intervening quality control. The inseparability of production and consumption of a service means that traditional service quality cannot be engineered at a manufacturing plant then delivered in its final form to the consumer. In many service industries, quality becomes evident at the time of delivery.

3.6.2.3 Evaluation of Service Quality

If quality is defined as the degree to which a service fulfils consumers' expectations, then the issue lies in recognising what these expectations are. The general absence of easily understood criteria for assessing quality makes the articulation of consumers' requirements, and the communication of the quality level on offer, much more difficult than is the case for goods (Palmer, 1998). As stated earlier, service quality is a highly abstract and elusive construct, in comparison to goods where technical aspects of quality take precedence. Therefore many conceptualisations address these abstract consumer expectations in respect of quality. Consumers on the other hand, judge service quality as the extent to which perceived service delivery matches up to initial expectations. By extension, a service which is perceived as being of mediocre standard may be considered of high quality when compared against low expectations, but of low quality when assessed against high expectations.

The analysis of service quality is further complicated by the fact that production and consumption for service generally occur simultaneously. Rathmell (1974) notes that there are two interfaces between the production and consumption of services. One is via the standard marketing mix and the other is through what Rathmell calls Buyer-Seller Interaction. The marketing mix is the set of tools available to an organisation to shape the nature of its offer to customers (Palmer, 1998). Recently the traditional 4Ps of the marketing mix, i.e. products, pricing, promotion and place, have been found to be too limited in their application to services, and have been extended to cover people, physical evidence, processes and customer service (Palmer, 1998).

Gronroos (1984) emphasises that buyers of manufactured goods only encounter the traditional marketing mix variables, i.e. the product, its price, its distribution and how these are communicated to them. Usually, production processes are unseen by consumers and therefore cannot be used as a basis for quality assessment. By contrast, service inseparability results in the production process being an important basis for assessing quality. A further problem in understanding and managing service quality flows from the intangibility, variability and inseparability of most services which result in a series of unique buyer/seller exchanges, with no two services being provided in exactly the same way.

How, then, should service quality be measured? Major research contributions have emanated from North America (Parasuraman, *et al.*, 1985, 1988, 1991 and 1994; Cronin and Taylor, 1992 and 1994; Day, 1984; Oliver, 1980); from Scandinavia (Gronroos, 1984 and 1988); and from the UK by Johnston, *et al.*, (1990) reviewed by Lewis, *et al.*, (1994). Major unresolved issues emerging from this ongoing debate include the empirical vs. diagnostic value of expectations in service quality measurement and the relative merits and demerits of SERVQUAL vs. SERVPERF formulations of the perceptions gap (Parasuraman, *et al.*, 1994).

Disconfirmation, Gap Theory and SERVQUAL

The *expectancy disconfirmation* pattern in the consumer satisfaction literature contends that consumer satisfaction results in large part from the disconfirmation of prior expectation (Oliver, 1980). To put it simply, if the performance of a service provider meets or exceeds expectations, then the consumer is more likely to be satisfied. If the performance fails to meet expectations, then the consumer is more likely to be dissatisfied.

Gap theory is a parallel research stream involving similar ideas. *Gap* theory suggests that the difference between consumers expectations about the performance of a general class of service providers and their assessment of the actual performance of a specific firm within that class drives the perception of service quality (Parasuraman, *et al.*, 1985). *Gap* theory was the forerunner of the *SERVQUAL* approach (Parasuraman, *et al.*, 1988, 1991). *SERVQUAL* is a 22-item scale for measuring service quality along five dimensions: reliability, responsiveness, assurance, empathy and tangibles (these five dimensions are recast from the 10 dimensions discussed under service quality characteristics). Each statement in the *SERVQUAL* questionnaire is one of two types. One measures expectations about firms in general within an industry and the other measures perceptions regarding the particular company whose service is being assessed. For example,

Service expectations:

'consumers should be able to trust bank employees'
'banks should have up-to-date equipment'

Strongly agree				Strongly disagree		
1	2	3	4	5	6	7

And perceptions

'you can trust the employees of xyz bank'
'xyz bank has up-to-date equipment'

Strongly agree				Strongly disagree		
1	2	3	4	5	6	7

SERVQUAL operationalised service quality by subtracting consumers' expectations scores from their perception scores on the 22 items.

While the *SERVQUAL* instrument has generated considerable interest in service quality measurement, several studies have pointed out the pitfalls with using the difference scores. Carman (1990) contends that little if any theoretical or empirical evidence supports the relevance of the expectations-performance gap as the basis for service quality. In particular, the need to measure expectations (for example, Babakus and Boller, 1992; Cronin and Taylor, 1992 and 1994), the interpretation and operationalisation of expectations (for example, Teas 1993 and 1994), the reliability and validity of SERVQUAL's difference score formulation (Babakus and Boller, 1992; Brown, *et al.*, 1993), and SERVQUAL's dimensionality (for example, Carman, 1990).

The *SERVQUAL* instrument involves two separate lists of statement, i.e. one for expectations, another for perceptions. This may lead to a situation, whereby respondents are unsure about the differences between the two lists of statements for the same item. Further, the meaning of the (i.e. normative) word 'should' is problematic and respondents may have difficulty in remembering to rate companies in general as opposed to what they want or expect from a particular company. As a result, constant reinforcement of the difference between the two sets of statements is needed. In addition, because all 22 expectation statements are presented prior to perception statements, respondents may forget how they have rated expectations when completing their performance statements.

All statements in the *SERVQUAL* questionnaire are treated as equally important. This may be inappropriate as determinants of service quality differ in their importance to individual respondents and in different service environments. For example, is the fact that employees are well dressed as important as the organisation being dependable? The *should* responses to the expectations questions provide an idea of the relative importance of an element, although traditional gap measurement is not weighted as a result of this information. Lewis and Mitchell (1990) suggest that it ought to be. Parasuraman, *et al.*, (1991) in their revision of the *SERVQUAL* instrument, propose an additional question: respondents are given 100 points to allocate between the five dimensions of quality and asked how important each of these is when evaluating a company's quality of service, such that the relative importance and gaps for the five dimensions may be established.

Some negatively worded statements, which were included in the original *SERVQUAL* to encourage respondents to read questions carefully, can be confusing to respondents and take them some time to work out the logic, for example,

'my bank manager does not take time to discuss my business'

Strongly agree				Strongly disagree		
1	2	3	4	5	6	7

This problem is exacerbated by the use of 44 scales in *SERVQUAL*, many of which appear to the respondent to be repeated, which may lead to loss of interest and completion errors. Consequently, replicated studies tend to avoid negative wording and Parasuraman, *et al.*, in their 1991 revision of the scale completely changed the negative statements into a positive format.

The *SERVQUAL* scale has no verbal labels for points 2 to 6 and so respondents may be over using the extremes of the scales which do have verbal anchors because subjects have a tendency to focus on words (Lewis *et al.*, 1994). A further issue of concern is that the 7-point Likert scale used in *SERVQUAL* may camouflage subtle variations in consumer expectations and perceptions. For example, a consumer may feel his expectations lie between points 4 and 5, and record 5 (rather than 4.4) and his performance rating between 5 and 6 (rather than 5.6), therefore the recorded measurements show no difference between expectations and perceptions when the actual difference is 1.2. These slight variations become increasingly important if one begins to weight the items according to their importance.

Additionally, little if any theoretical or empirical evidence supports the relevance of the expectations-performance gap as the basis for measuring service quality (Carman, 1990). This has led many authors to recommend the direct measurement of the disconfirmation (Babakus and Boller, 1992; Carman, 1990). The direct measurement of disconfirmation results in a scale of the type 'much better than expected' to 'much worse than expected'.

Perceived Service Quality (SERVPERF)

Due to the difficulties and uncertainties surrounding the definition of service quality, some of the marketing literature appears to offer considerable support for the superiority of simple performance-based measures of service quality (Bolton and Drew, 1991; Churchill and Surprenant, 1982; Cronin and Taylor, 1992).

As noted above in the definition of service quality, if service quality is considered to be an attitude, Oliver (1980) argues that this implies several outcomes. Firstly, in the absence of prior experience with a service provider, expectations initially define the level of perceived service quality. Secondly, upon the first experience with the service provider, the disconfirmation process leads to a revision in the initial level of perceived service quality. Thirdly, subsequent experiences with the service provider will lead to further disconfirmation, which again modifies the level of perceived service quality. And finally, the redefined level of perceived service quality similarly modifies a consumer's purchase intentions toward that service provider. What this means is that service quality and consumer satisfaction are distinct constructs, but are related in that satisfaction mediates the effect of prior-period perceptions of service quality to cause a revised service quality perception to be formed. This logic questions the disconfirmation framework as the primary measure of service quality, because disconfirmation appears only to mediate, not define, consumers' perceptions of service quality (Cronin and Taylor, 1992).

Churchill and Surprenant (1982), conducted two experiments to examine the effects of expectations, performance, and disconfirmation on satisfaction. The results of one experiment suggested that performance alone determines the satisfaction of subjects. However, Woodruff, *et al.*, (1983), using the *adequacy-importance* model, indicate that assimilation/contrast theory suggests that consumers may raise or lower their performance beliefs on the basis of how closely perceived performance approximates expected performance. They suggest that including importance weights and expectations only introduces redundancy, thus contributing additional support for performance only measures of attitude. Bolton and Drew (1991b), based on a field experiment, conclude that current performance ratings strongly affect attitudes whereas the effects of disconfirmation are generally insignificant and transitory. This study is particularly significant because the attitude examined is consumers' perceptions of the quality inherent in a service. Cronin and Taylor (1992), based on a field experiment of four service industries, concludes that the performance only approach, which they named *SERVPERF,* is superior to the *SERVQUAL* approach, in measuring service quality.

Based on the above studies, it may be concluded from the satisfaction and attitude literature that; firstly, perceived service quality is best conceptualised as an attitude; secondly, the *adequacy-importance* model is the most effective *attitude-based* operationalisation of service quality; thirdly, current performance adequately captures consumers' perceptions of service quality offered by a specific service provider; fourthly, service quality is an antecedent of consumer satisfaction and consumer satisfaction exerts a stronger influence on purchase intentions than does service quality. In addition to the theoretical support for the performance-based measures of service quality, practitioners often measure the determinants of overall satisfaction/perceived quality by having consumers simply assess the performance of the company's business process (Cronin and Taylor, 1992). Cronin and Taylor (1994), contend that judgements of service quality, and satisfaction appear to follow the evaluation of a service provider's performance.

This extended discussion on the service quality construct and competing theories on its measurement highlight the importance of the influence of service quality in influencing consumer purchase intentions. As it has been already described in section (3.6.1) of this chapter, and examined in further detail in section (4.5), Internet banking services display characteristics that are more akin to physical products as opposed to service products. It follows from this that many service quality dimensions in Internet banking services can be evaluated using objective criteria. Therefore, on balance, these characteristics favour the evaluation of the performance of the service or the SERVPERF approach to the measurement of service quality in Internet banking services. These conclusions drawn here are important in the formulation of hypotheses in chapter five, developing data collection methodologies m chapter six and the operationalisation of testing hypotheses in chapter seven.

3.7 A Critique on Interaction of Service Quality, Consumer Satisfaction and Purchase Intentions

The preponderance of evidence in the existing literature on the marketing of services has identified the critical roles of service quality and satisfaction in the formation of consumers' purchase intentions. However, despite the great strides made in recent years, our understanding of the specific nature of the relationship between service quality and consumer satisfaction, as well as how these two constructs combine to impact consumer purchase intentions, continues to perplex marketing scholars (Gronroos 1993; Rust and Oliver 1994).

The weight of the evidence in the services literature reviewed above supports the position that service quality and consumer satisfaction are best conceptualised as unique constructs that should not be treated as equivalents in models of consumer decision making. The conclusion that service quality and satisfaction are unique constructs naturally begs the question, "What then is the specific nature of the relationship between these constructs in the formation of consumers' purchase intentions?"

The relationship between purchase intentions and consumer satisfaction has been addressed in several studies, including Bearden and Teel (1983), Oliver (1980), and Oliver and Swan (1989). LaBarbera and Mazursky (1983) investigate the relationship between actual purchase behaviour and consumer satisfaction. There have also been a number of recent empirical attempts to validate the specific nature of the relationship between service quality and consumer satisfaction in the formation of consumers' purchase intentions (Bitner, 1990; Cronin and Taylor, 1992; Woodside *et al.*, 1989). Unfortunately, the evidence to date has demonstrated conflicting results. For example, Woodside *et al.*, (1989) proposed one of the first models specifically assessing the relationships between service quality perceptions, consumer satisfaction judgements, and behavioural intentions in the marketing literature and report empirical results suggesting that consumer satisfaction is an intervening variable that mediates the relationship between service quality judgements and purchase intentions i.e.,

service quality → satisfaction → purchase intentions.

Bitner (1990) investigated the service quality and consumer satisfaction perceptions of 145 travellers at an international airport. The basis of Bitner's model is an attempt to reconcile Oliver's (1980) disconfirmation of expectations *paradigm with attribution* theory. Bitner hypothesises that consumer attributions mediate disconfirmation and satisfaction judgements. Service quality judgements, in tum, are hypothesised to mediate consumer satisfaction judgements and consumers' behavioural intentions. Thus, Bitner suggests an alternative ordering of the service quality and satisfaction constructs, i.e.,

satisfaction → service quality → behavioural intentions.

Bitner's fully recursive path analysis results appear to support her hypothesised model and thus contradict the causal order reported by Woodside *et al.*, (1989).

Cronin and Taylor (1992), in a non-recursive structural model, present the first simultaneous test of both of the aforementioned relationships. Although Cronin and Taylor hypothesise a priori that satisfaction is an antecedent of service quality, the empirical results of their LISREL-based analyses indicate the opposite. In addition to being the first to simultaneously assess both of the possible relationships between service quality and satisfaction, Cronin and Taylor also are the first to use a multi-industry sample in their investigation of these research questions. Specifically, these authors directly assess the service quality/consumer satisfaction relationship across four industries: banking, pest control, dry cleaning, and fast food. For each of the four service industries they investigated, Cronin and Taylor's results support Woodside *et al. 's,* (1989) conclusion that service quality appears to be a causal antecedent of consumer satisfaction.

Thus, the results of empirical efforts to validate the specific nature of the relationship between service quality, consumer satisfaction, and purchase intentions have supported both possible relationships between the constructs, i.e.,

> service quality → satisfaction → purchase intentions, satisfaction → service quality → purchase intentions.

Hence, the literature review above identifies the conflicting nature of the empirical results to date. The conceptual nature of the relationship is perhaps best explained by Rust and Oliver (1994). These authors take the perspective of Cronin and Taylor (1992) that satisfaction is subordinate to quality - that is, quality is only one of many potential service dimensions factored into consumer satisfaction constructs. However, these authors continue by suggesting that satisfaction may also indirectly reinforce quality perceptions.

Perhaps more importantly, however, recent conceptualisations of the relationship between service quality and consumer satisfaction, that form the basis for Rust and Oliver's (1994) arguments, have incorporated the additional dimension of level of aggregation. In short, a number of recent authors have supported the position that both service quality and consumer satisfaction exhibit both global and encounter-specific forms (Anderson and Fornell, 1994; Bitner and Hubbert, 1994; Rust and Oliver, 1994).

The conflicting empirical results reported in the literature are to some extent due to the fact that the models reported have been constrained to tests of the main effects of service quality and satisfaction on purchase intentions.

While the relationships between these constructs is not the main emphasis of this thesis, this extended discussion is crucial for it is necessary to explain how and why these constructs are important in influencing consumer purchase intentions. The critique reiterates, firstly, the importance of service quality in influencing purchase intentions in service products, and secondly, the interaction between these constructs. Having examined both arguments, the stance taken in this thesis is that service quality is an antecedent of consumer satisfaction, which in tum exert a strong influence on purchase intentions. This is a very important conclusion in that it forms a central tenant in the formulation of the hypotheses, which are discussed in chapter five.

3.8 Chapter Summary

Service products have been recognised as a distinctive product category and that, as consumer behaviour is affected by product, the consumption of services merits consideration in its own right. There are considerable problems with trying to apply a service process driven approach to the consumption of services since a number of the 'stages' in the decision model occur simultaneously or in reverse when dealing with a service product. For instance, physical goods are evaluated, purchased and then consumed, whereas services are evaluated, purchased and consumed at the same time.

The implications for consumers in the pre-consumption phase of services products are investigated. Drawing upon the body of knowledge which explains consumer behaviour in relation to goods, this suggests problems for consumers in making consumption decisions in services. The nature of the consumers' experience with the service encounter revealed seven service dimensions. The process of searching for information, learning about the service offerings and evaluating different services provides a context for the encounter embodying expectations and learned responses. Equally, how the encounter was experienced and cognitive responses, to people and service, will impact upon how the service organisation is reviewed. While the separation of the encounter into a discrete process serves both analytical and conceptual facility.

Chapter 4.

From Financial Services to Internet Banking Services

4.1 Introduction

The discussion in the previous chapter demonstrates that many individual factors influence consumer purchase decisions, and that these factors can be viewed from a number of different theoretical perspectives. From the many models of buyer behaviour reviewed, it was argued that the cognitive consumer behaviour offers the best framework for analysis. In this analysis it seen that consumers face a number of problems as buyers of services. The objectives of this chapter are; firstly, to examine these aspects in more detail, and thereby help in the understanding of consumer purchase intentions in Internet banking services, and secondly, to explore the qualities of Internet banking services in detail.

It does so by reviewing literature on consumer purchase of financial services followed by an analysis of previous research in consumer buying behaviour in financial services. Ennew and McKechnie (1998) argue that consumers face far greater problems when buying financial services, and that these services are often highly complex and may be difficult to evaluate at both pre and post purchase. They also note that due to these additional problems the understanding of the process is still imperfect. Some of these difficulties were also discussed briefly in the previous chapter. The third section describes the Internet environment, tracing the development of the Internet, the World Wide Web and attributes of Web sites. The final section of the chapter describes the qualities of Internet banking services by drawing upon the body of literature reviewed in this chapter and also by developing upon arguments presented in previous chapters. The chapter concludes with a summary of the key issues.

4.2 Consumer Behaviour in Traditional Financial Services

The concept of the cognitive consumer adopted in the previous chapter is based on a fundamental agreement that the actual purchase process is a discrete but interlinked sequence of tasks characterised as a problem solving exercise. The consumer decision process begins with a needs recognition, or what some authors refer to as the "problem" (Ennew and McKechnie, 1998). To solve this problem the buyer goes through a process of information acquisition so that available alternatives can be identified. Once information has been assembled about the financial service product in question, a decision is made on the purchase, based on an evaluation of the available alternatives. It was also seen that although this conceptualisation is not devoid of weaknesses, such as the assumption of a very high degree of rationality on the part of the consumer, it offers a clear framework on which to base the discussion and to consider the difficulties that consumers confront when buying financial services. One of the reasons for this inaction is that needs of this nature are uninteresting and are inherently complex.

4.2.1 Problem Recognition

The first stage of the purchase process is the increasing recognition of needs that can be fulfilled through a purchase of financial services, including the need to make payments (cheques, debit cards, etc.), the need to defer payment (loans, mortgages, credit cards, etc.), the need for protection (insurance, etc.), the need to accumulate wealth (unit trusts, ISA's, etc.) and the need for advice (tax, financial planning, etc.) (Ennew and McKechnie, 1998). An important distinction here is in the recognition that not all needs will be translated into goal-oriented behaviour (Gabbot and Hogg, 1998), i.e. an individual may recognise a desire to apply for a particular credit card, but not be motivated to do anything about it. One of the reasons for this inaction may be that needs of this nature is uninteresting and is inherently complex.

4.2.2 Information Acquisition

Consumers may acquire information about alternatives from memory, advertisements, other people, trade magazines, company databases, and, last but not least, from the Internet. However, if the consumer is uninterested (either because the product is complex or as highlighted earlier financial products may not seen to be "attractive" purchases) in the purchase it induces a certain level of passivity and this may affect the extent of the information search (Ennew and McKechnie, 1998).

Zeithaml (1981) suggested a framework for information gathering based on *search qualities,* however financial services are low in search qualities due to their intangibility and inseparability. The consumer may overcome this weakness by relating to *experience qualities,* these are attributes that can only be determined after purchase or during consumption. Experience can come from any of the following sources: (1) the consumer could have used the service provider before for a similar service experience, (2) a pre-purchase trial, (3) observation of others receiving the service experience, (4) reliance upon the experiences of others (Locander and Hermann, 1979). Own experience is problematical, since most financial services are long term, continuous or both. Then there will be a tendency to rely on other sources of experience.

Apart from the above problems, Ennew and McKechnie (1998) note that consumers face further problems in relation to the validity and accessibility of information. Firstly, many financial services are long term in nature; therefore even if consumers gain explicit experience from word-of-mouth recommendations, that experience may be at best partial since the full benefits of a product (say a 'with-profits' bond for example) may not have been realised. Secondly, a large number of financial services products tend to be customised to individuals (reflecting marriage status, age, etc.), drawing on the experience of others may be misleading if personal circumstances are different. Thirdly, the complexity of many financial services means that many consumers may collect information, but not actually understand the information or interpret it incorrectly.

4.2.3 Evaluation of Alternatives

The previous paragraphs highlighted the difficulties consumers face with respect to information gathering. These difficulties tend to be amplified when the consumer attempts to evaluate alternatives. The reliance on the experience qualities of financial services by consumers makes pre-purchase evaluation difficult. In instances where credence qualities are significant, as the very definition of credence qualities suggests (Zeithaml, 1981), post-purchase evaluation is problematical (Ennew and McKechnie, 1998). In addition, the availability of a variety of different products which may satisfy a particular need compounds the problem; for example, the consumer who wishes to set aside savings for the future retirement may consider a range of products, such as, pension plans, national savings certificates, bonds, unit trusts, equity investments in quoted companies, etc. The risk-return characteristics of these services vary considerably, as a consequence this is reflected in prices, and there is rarely any easy way to make direct comparisons across different service product types. These problems have been aggravated by a lack of transparency in the pricing and promotion of many financial services (Diacon and Ennew, 1996).

Just as the presence of credence qualities in many traditional financial services make the information acquisition and search process difficult, it also makes the evaluation process complex. Products which contain a significant element of advice or which require active management over the course of their life may be difficult to evaluate even after purchase (Ennew and McKechnie, 1998). The performance of many long-term investment products, such as unit trusts, pension funds, investment trusts, etc. is determined in part by the skill of the relevant fund managers and in part by economic factors, which are beyond the control of the supplier. Therefore, consumers who purchase these service products, expose themselves to certain risks (both actual and perceived). However, when it comes to assessing the performance of these service products, they will experience difficulties in determining whether poor performance was due to company specific factors or to external contingencies. As cited earlier, a consequence of this situation is a tendency for customers to evaluate service providers (rather than the services themselves) and to rely heavily on trustworthiness as an attribute of those providers.

This review concentrating on the pre-purchase stage of a financial service has to a large extent mirrored the discussion in section (3.5).

4.2.4 Purchase and Post-Purchase Evaluation

Once consumers complete the evaluation of alternatives, they may make a purchase. Consumer interaction with traditional financial service provider staff is of significance in the purchase process, for such interaction is an integral part of the service encounter experience. As the discussion at the problem recognition stage shows, it is possible that with many financial services consumers, the need for financial service products may be created or activated at the point of purchase (Ennew and McKechnie, 1998). Consequently, the actual purchase process will often be the result of an active selling effort by a supplier (Ennew and McKechnie, 1998). In this case customer interaction with financial service provider staff is likely to be of even greater significance in the purchase process.

The purchase process is influenced by the inseparability of production and consumption in traditional financial services. Customer participation is essential for a service encounter to take place. Further the influence of other dimensions of a service encounter, i.e. that of the service context, proximity, duration and customisation on purchase intentions were discussed at length in section (3.6). Therefore it is fair to conclude that the quality of the service output very much depends on the nature of the personal interactions between the customer and the financial service provider staff.

Fiduciary responsibility is often highlighted as an important characteristic that distinguishes financial services from other services and goods

(McKechnie, 1992). Ennew and McKechnie (1998) note that one dimension of fiduciary responsibility is that suppliers need to exercise discretion with respect to the sale of certain products. Thus, for example, it would be inappropriate for a bank to lend money to a business which had few prospects for survival and success. However, until a consumer has signalled an intent to purchase it may not be possible to identify whether or not it is appropriate to provide that product to that customer. Thus, the consumer effectively faces the added problem that even if a conscious decision to purchase has been taken, the financial institution concerned may be unwilling to provide the product.

4.3 Previous Research on Consumer Buying Behaviour in Traditional Financial Services

The previous section highlighted some of the difficulties that consumers encounter in the purchase of traditional financial services. In this section some of the empirical research relating to buyer behaviour is examined to assess the degree to which it corroborates the issues discussed in the previous section.

McKechnie (1992) notes that, while in the context of physical goods consumer buying behaviour is supported by a large volume of empirical work, as far as services marketing and the financial services sector in particular is concerned, the empirical work is not well developed. She gives three possible reasons for this, (1) there may be a problem with the conceptual models themselves in that they do not lend themselves to empirical testing, (2) it is not clear whether these models are necessarily the most appropriate conceptual frameworks to use in any case, and (3) there has been a lack of appropriate measures of salient dimensions for testing concepts in services marketing situations.

In spite of weaknesses in the theoretical framework, there have been a number of studies of buying behaviour of financial services. Ennew and McKechnie (1998) summarise studies up to 1998. A central feature of these studies (supplemented here by later studies) is that they have not attempted to test whole conceptual framework. Instead, the focus has been on specific issues in relation to purchase behaviour such as factors affecting the choice of bank and usage of financial services, consumer loyalty, consumer expectations and perceptions, and service quality. The results of the main studies are summarised in Table (4.1, next page).

Factor(s) influencing purchase decisions	Author(s)
Pre-purchase stage of financial services	
Size of bank	Leonard and Spencer, 1991; Arora, *et al.*, 1985
Economic rationale (price), competitive interest rates	Leonard and Spencer, 1991; Jain, *et al.*, 1987
Locational convenience	Laroche, *et al.*, 1986; Martenson, 1985; Khazeth and Decker, 1992
Branding, dependability	Arora, *et al.*, 1985; Laroche and Manning, 1984; Devlin *et al.*, 1995; Khazeth and Decker, 1992
Variety of services, loan availability	Leonard and Spencer, 1991; Martenson, 1985; Arora, *et al.*, 1985; Khazeth and Decker, 1992
Consumer loyalty	Jain, *et al.*, 1987; Bloemer, *et al.*, 1998
Purchase and Post-purchase evaluation of financial services	
Service quality	Lewis, 1991; Lewis, *et al.*, 1994; Newman and Cowling, 1996; Newman, *et al.*, 1998; Avkiran, 1994 and 1999; Levesque and McDougall, 1996; Athanassopoulos, 1997; Newman, *et al.*, 1998; Bloemer, *et al.*, 1998; Khazeth and Decker, 1992
Human aspects including friendliness, courtesy, competence	Leonard and Spencer, 1991; Jain, *et al.*, 1987; Laroche, *et al.*, 1986; Lewis, *et al.*, 1994; Newman, *et al.*, 1998; Avkiran, 1994 and 1999; Newman, *et al.*, 1998
Speed of service	Laroche, *et al.*, 1986; Khazeth and Decker, 1992
Nature of delivery	Furlong and Ritchie, 1986; Athanassopoulos, 1997
Ease of transactions	Arora, *et al.*, 1985

Table (4.1): Factors influencing purchase intentions in traditional financial services

On the whole, these empirical studies highlight the importance of factors such as confidence, trust and consumer loyalty. Some of the common criteria in bank selection are dependability and size of the institution, location, convenience, ease of transactions, service quality and human aspects of bank personnel. It would appear that consumers are more interested in how the service is delivered (i.e. the functional quality dimension) rather than what is actually delivered (i.e. the technical quality dimension). McKechnie (1992) observes that this is not surprising given the difficulties consumers have in evaluating services, which were highlighted in the previous section.

4.3.1 Service Quality in the Traditional Financial Services Sector

In most models of consumer evaluations of financial services, the focus has been on a comparative judgement of expectations versus perceived performance resulting in the two major evaluative judgements of perceived service quality and consumer satisfaction.

Both concepts have been frequently used and measured m the financial services. Consumers form expectations prior to their encounter with a financial service organisation, they develop perceptions during the service delivery process and subsequently they compare their perceptions to their expectations in evaluating the outcome of the service encounter. While service quality and satisfaction are concepts that have a number of similar characteristics, they have points of differentiation as well, as illustrated in the critique of these constructs in the previous chapter.

A large number of studies summarised in Table (3.1), examine the relationship between purchase intentions of financial services and service quality, and the measurement of service quality in financial services, emphasise the roles played by both technical and functional dimensions of the service. A number of these studies have also found that the performance of the delivery function of a financial service has an impact on service quality perceptions. Additionally, a common theme of all these studies is the important part played by human contact. Avkiran (1994 and 1999) in particular emphasise that, staff conduct as the key variable explaining the direct and indirect linkages to service quality.

4.3.2 Price, Perceived Value and Brands in Traditional Financial Services

There is indeed very limited published research on pricing in the financial services sector. Among the most important studies on pricing in financial services, are Leonard and Spencer (1991), who discuss the importance of conveying the perception of value in a bank's prices. They suggest that this could be done through creating the 'right' pricing environment through research, pricing committees and employee-training programmes. They also examine the variables influencing pricing in banks, such as operating costs, elasticity of demand and market segmentation.

Bharadwaj *et al.*, (1993) argue that when buyers cannot easily evaluate the qualities and value of the service or capabilities of the service provider, then brand reputation may serve as an important proxy for more detailed evaluations. In the case of financial services, branding is often at the organisational level (Devlin *et al.*, 1995), lending further support to the notion that organisation-wide factors will be important in adding value. Bharadwaj *et al.*, (1993) go further and state that those services which are highly intangible and are, therefore, likely to be high in experience and credence qualities, will find brand reputation important as a potential competitive advantage.

Qualitative research focusing on financial services offerings (Devlin and Ennew, 1997) show that attempts to add value and achieve competitive advantage may be contingent upon, (1) the complexity of the service offering in question and (2) the perceived knowledge and sophistication of the target market. For certain, more complex financial services offerings in particular, their high intangibility makes it potentially difficult to envisage the use of specific services features as the basis for establishing a competitive advantage, due to lack of consumer cognition.

A lack of consumer understanding may result in increased reliance on experience and credence qualities of the financial service offerings, which may, in turn, enhance the importance of organisation-wide factors such as image reputation, and quality. However, as identified earlier, this may be less of a problem in the case of financial services with features and benefits which consumers find relatively easy to understand and evaluate.

In contrast with traditional financial services, Internet banking services represent a new delivery method devoid of human participation at face-to-face level, innovative financial products and a theoretically frictionless market. Consumers would experience different service characteristics in this new environment. Hence it is now opportune to begin to investigate these aspects. A brief introduction to the Internet and Internet banking services were made in section (1.2). The following sections will examine the Internet in greater detail, and illustrate attributes of the Internet in general and nature of Internet banking services in particular.

4.4 The Internet

At this stage, it is useful to examine the development of the Internet from its formative stages, for it would assist in the understanding of the nature of Internet banking services and therefore the research question.

4.4.1 The Development of the Internet

The beginning of the Internet can be traced to 1960s and the year 1969 stands out in particular, when the Advanced Research Projects Agency Network (ARPAnet) was commissioned by the Department of Defence (DoD) of the United States of America for research into networking (Zakon, 1998), linking the military with defence contractors and universities carrying out defence research. Only certain activities were permitted, while others were forbidden. The type of activities that were 'allowed' were enshrined in the National Science Foundation's (NSF) Appropriate Use Policy (AUP), which for years inhibited commercial activity on the Internet (Dem, 1994).

By late 1980s several networks were in operation. These included PSI, Alter-et, CERNet, BARRNET, CICNet, Neamet, PREPnet, etc. Although there were so many networks in operation at that time, the Internet as we know it today was not a reality. The seeds to the present day interconnected networks were first sown on March 25, 1991, when PSI, Alter-Net and CERFnet announced the formation of the Commercial Internet eXchange (CIX) (Dem, 1994). This was an interconnection point for Internet member networks. Belonging to CIX meant a network, and by extension the network's member sites and their member users could reach other CIX members without having to involve NSFNET, and hence the AUP did not apply. From this point onwards commercial activity proliferated.

The NSF was gradually "privatised", and this process concluded in April 1995, with the de-funding of the NSFNET backbone. The proceeds of this exercise were distributed to regional networks to buy national scale Internet connectivity from now numerous, private, long haul networks (Leiner, *et al.*, 1997)

4.4.2 The World Wide Web (WWW)

In the present day context, the bulk of consumer Internet traffic can be attributed to the WWW (Sagawa, 1997). The WWW is a body of computer files conforming to Hyper Text Markup Language (HTML) available on host computers connected to the Internet. The HTML standard allows non expert computer users easily to access multimedia content (text, graphics, audio, video, and so on) using commonly available browser software. The WWW was developed to be a pool of human knowledge, which would allow collaborators in remote sites to share their ideas and all aspects of a common project (Bemers-Lee, *et al.*, 1994). The property of scaling has allowed the Web to expand rapidly from its origin at the Centre for European Academic Research Network (CERN) across the Internet irrespective of boundaries of nations or disciplines.

The basis of the WWW can be traced to a notebook programme written by Tim Bemers-Lee in 1980, called "Enquire-Within-Upon Everything" which allowed links to be made between arbitrary notes. Between March 1989 and May 1990, the same author circulated a proposals on Information management and HyperText, and as a result, by November 1990 an initial WWW program was developed. The development quickened, by Christmas 1990 a line mode browser was demonstrable at CERN. However it was not until late 1993 that "Mosaic" proliferated and the use of WWW for commercial purposes took off (Zakon, 1998). By 1994, WWW edged out telnet to become the second most popular service on the Internet (behind ftp-data), based on the percentage of packets and bytes traffic distribution on NSFnet. The following year WWW became the service with the greatest amount traffic on the Internet, and continues to grow in the present day.

The above narrative account illustrates the development of Internet infrastructure and the Web, the most visible form of traffic on the Internet. The next questions that needs addressing are, **(1)** how the Internet environment is best examined in an academic context?, and (2) how can consumer behaviour in this environment be analysed?

4.4.3 The Internet Environment

In examining the Internet environment, it is useful to consider the Internet as a unique hypermedia Computer Mediated Environment (CME). Hoffman and Novak (1996b and 1996c) define a hypermedia CME as a dynamic distributed network, potentially global in scope, together with associated hardware and software for accessing the network, which enables consumers and firms to (1) provide and interactively access hypermedia content (i.e. machine interactivity) and (2) communicate through the medium (i.e. person interactivity). Navigation is defined as self-directed movement through a hypermedia CME (Hoffman and Novak, 1996b). The WWW is the most visible component of the Internet, and it is the primary medium through which Internet banking services are marketed and delivered. It is considered to be a many-to-many communication model that turns traditional principles of mass media advertising (based on the one-to-many model) inside out. Many-to-many communication takes place in a distributed network, where the interactivity can be with the medium (i.e. machine interactivity) in addition throughout the medium (i.e. person interactivity) (Hoffman and Novak, 1996b). On the other hand, one-to-many communication is a typical mass communication model, whereby there is no interaction of transmitted content with the consumers (Hoffman and Novak, 1996b). This renders impossible the blind application of marketing and advertising approaches that assume a passive, captive consumer (Hoffman and Novak, 1996b).

4.4.3.1 Consumer and the Internet Environment

As stated above, the next question that needs examining is how do consumers behave in this unique hypermedia Computer Mediated Environment?

Consumer behaviour in this environment can be either functional and goal directed or hedonic search behaviour. Web users of functional goal directed behaviour are looking for specific information on the Website in order to accomplish a certain goal (Muylle, *et al.*, 1998). The search is a means to an end. Users displaying hedonic search behaviour want to be entertained and they consider the Web to an end in itself. Consumers can actively choose to approach banks through their Web sites and exercise unprecedented control over the management of the content with which they interact. Consumers are able to decide for themselves when they want to access the site at their own convenience.

Consumers can vary in their ability to achieve flow. Hoffman and Novak (1996b) define flow experience in a CME as the state occurring during network navigation, which is (I) characterised by a seamless sequence of responses facilitated by machine interactivity, (2) intrinsically enjoyable, (3) accompanied by a loss of self-consciousness, and (4) self-reinforcing. Therefore, it is clear that depending on their motivation (i.e. the level of attention to the interaction) and skills, consumers can be segmented. The implication for service providers on the Internet, such as Internet banking services, is that they can offer services in a more focused manner to the desired consumer segments. Visits to a Web site are increased if the environment facilitates the flow state. This is especially important since a consumer may only simulate a purchase process on the first visit, and consequently if the Web environment facilitates the flow state it will positively influence purchase intentions by inducing further visits. Further, the purchase of an Internet banking service may not be a one-off purchase, and in order to ensure that consumers create long term banking relationships an environment that facilitates the flow state is of paramount importance.

4.5 The Qualities of Internet Banking Services

So far, this chapter has examined consumer behaviour in relation to financial services on a theoretical level and previous empirical research in financial services sector. This has highlighted some of the problems that confront consumers when choosing financial services. These difficulties are due partly to the generic qualities of services, partly to unique features of financial services and partly to the practises employed within the industry itself (Ennew and McKechnie, 1998). This section will draw from key qualities of traditional financial services; intangibility, inseparability, heterogeneity, fiduciary responsibility, two-way information nature and the long-term nature of many of the products to illustrate the qualities of Internet banking services. This exercise is important for two reasons. Firstly, it brings together concluding arguments of the analysis carried out in previous chapters and present chapter of this thesis. Secondly, it is very important to develop a framework to examine Internet banking services in order to address the research question, since the qualities of Internet banking services influence consumer purchase intentions.

Banking services can be located at the "pure service" extreme of the goods-services continuum (Heaney and Goldsmith, 1999). Hence, banking services can be defined as intangible products that are labour or skill intensive, and tangible products such as computers and buildings are used mainly to support this provision of labour. While Internet banking services contain these essential generic qualities of banking services to some extent, they do deviate significantly in their detail.

For the purposes of this thesis, as stated in section (1.2.2), the term Internet banking services is restricted to current and savings accounts offered to personal consumers. The analysis begins with a reconsideration of the reconsideration of the qualities of services in general.

4.5.1 Intangibility

Intangibility, the absence of physical dimensions, is probably the dominant characteristic of banking, and Internet banking is no exception. It was also seen that this makes the perception of a service highly subjective and an abstract concept for the consumer. Although this inability to inspect or touch services means that in any pre-purchase situation other services, such as a haircut at a salon, remains abstract until has been consumed, banking service products on the other hand has some clearly defined specifications. It is the delivery process of banking service products that are high in attributes that can only be assessed after purchase or during consumption, and hence are high in experience qualities. By and large, Internet banking product specifications are either equal or superior to those of traditional banking service products. However in Internet banking services, the delivery process is automated, hence it is largely a homogeneous process (as will be discussed below). Further, with the increasing sophistication of the Internet, it is becoming possible to assess the service delivery attributes before the actual purchase. It then follows that, unlike other services, the inspection qualities of an Internet banking services are high.

To mitigate the problems associated with intangibility, traditional banks have sought to 'tangibilise' the service (Ennew and McKechnie, 1998). The provision of some physical evidence (whether essential or peripheral) is a common approach to dealing with intangibility (Shostack, 1982). Examples of peripheral physical evidence would include chequebook covers, promotional free gifts, etc. Essential physical evidence is typically associated with branch networks or head offices with the appearance and layout being used to give a tangible representation of the organisation. Internet banks on the other hand, appear to be operating in a continuum; on one extreme new Internet bank entrants are keen to highlight the 'virtual' nature of their operation, on the other hand, the traditional banks that have Internet banking services have the qualities described above.

Often physical evidence of this nature is supported by the use of a tangible image or name. Thus, for example, the *Northern Rock* is an organisational name that tries to link to an image of stability and security. Equally a tangible image or association such as the Black Horse (Lloyds) can serve a similar purpose (Ennew and McKechnie, 1998). The new entrants into Internet banking emphasise the 'novelty' of their operations by selecting quirky names such as 'egg' (Prudential), 'smile' (Cooperative Bank), 'IF' (Halifax), 'Cahoot' (Abby National), 'Marbles' (HFC), etc. which set them apart from the traditional banks.

Attempts to tangibilise a banking service seek to address the problem of lack of physical form. However this strategy is less effective in relation to product complexity and lack of consumer passivity, which as discussed earlier. Two key approaches are important in this respect. Firstly, to address the complexity issue there is a need to focus on reducing perceived risk through building trust and confidence; if the consumer cannot fully understand the nature of the service then they must be able to trust a supplier and to feel confident that their finances are being safely managed. Attempts to build such trust often rely on the longevity of the organisation. In complete contrast to the terrestrial world, as seen above, a number of Internet banks have made an effort to portray an image different to their parentage. Secondly, consumer passivity or lack of consumer interest in banking services can often be gained from the purchases of the product (Ennew and McKechnie, 1998). For example, an interactive calculator on the egg Internet banking Website allows the consumer to key in a value that they are prepared to save per month, and the duration for which they are willing to do it. The promotional wording on the Web page tends to emphasise the purchases which can be made as a result of the saving (whether cars, home improvements, holidays, etc.).

4.5.2 Inseperability

A financial service transaction, is a performance where the production and consumption aspects take place in real time, hence is *inseparable* (Bell, 1981; Gronroos, 1978). This is one obvious consequence of intangibility; and also means that consumers have considerable difficulties with respect to pre-purchase evaluation. An additional feature of financial services is that it is essentially a people-based service, except in a situation where a customer would use a machine such as an ATM. The implication of this inseparability and the high level of interaction between the service provider and consumer therefore leads to a lack of standardisation since both the purchaser and the individual delivering the service can alter both the way in which the service is delivered, as well as what is delivered. This process has important implications for evaluation (Gabbot and Hogg, 1998). One method of overcoming this difficulty in pre-purchase evaluation is to evaluate the organisation and to draw on the experience of others. Accordingly a common theme in the marketing of financial services is to emphasise the performance and quality of the organisation and its people in order that there will be a radiating effect from organisation to service product (Ennew and McKechnie, 1998). Such approaches may be reinforced by active attempts to secure word-of-mouth recommendations.

Internet banking differs from other financial services on this aspect perhaps to a higher degree than any other characteristic. The "production" of the banking service is carried out by a series of components, comprising of a central computer (a host computer) and all the peripherals connected to the customer's computer by a communications link. While it is true that the hardware and the software for this process has been assembled and maintained by the Internet banking service provider's staff, the banking service production and delivery process takes place depending in a large part, on the commands of the customer and far less than in the terrestrial world on the Internet banking service provider. Hence there is a clear separation between the parties. An analogy can be drawn with a department store or a park. The bank provides paths, routes and different services but the customer is free to wonder anywhere they choose, including leaving the Web site.

The implication of this separation is that the service delivered is in large part standardised. The determining factors that would alter this standard is limited to customer behaviour, quality of communication link and equipment used. It depends on customer behaviour because, as described above, the service responds to the commands of the customer. Any variation on this behaviour (for example a customer clicking on an active hyperlink not necessary to the execution may take them elsewhere) will disturb the standard service product. Perhaps the quality of the communication link and the specification of the computer used for Internet banking services affect the service outcome more than anything else. For all intents and purposes the time taken for service execution will depend on the specification and the amount of Web traffic. In extreme situations a customer might not be able to execute a service transaction at all.

4.5.3 Perishability

A related feature of inseparability of a financial service offer is its *perishability*. Unlike most goods manufacturers who can normally hold stocks to meet fluctuating levels of demand, services cannot be stored (Palmer, 1998), i.e. the financial services provider cannot stockpile services against a possible need to in the future - a service is only delivered when a customer demands it. The perishability of services reiterates the real time nature of the product described above.

In certain circumstances the consumer may decide to delay consumption but not to consume more in advance or requirements. For the purchaser of financial services the time at which they choose to use the service may be crucial to its performance and therefore to the consumer's experience. Further, the presence of other consumers can influence the service outcome, for example during the lunch hour, a bank teller might be highly stressed at having to service a long queue of customers all wanting to conclude the service transaction in the shortest possible time, and even a slightest anomaly can dramatically influence the service experience.

Internet banking service offerings, however, can be 'stored' without the burden of an incremental cost structure, until it is convenient for consumers to use them. As described above, in Internet banking the service is delivered when the customer demands it, and the they are in control of many aspects of the service experience. Significant customer demand at peak times can create bottlenecks in system capacity and response times. These can be avoided by installing hardware systems with higher capacities. However it is appreciated that in most cases the communications link is not within the control of the Internet banking service provider, hence any bottlenecks in the communications system will inevitably affect the service experience.

4.5.4 Heterogeneity

Heterogeneity of financial services is related to inseparability and perishability. Traditional financial services are produced and delivered by individuals and therefore each service encounter will be variable by virtue of the participants, the time of performance or the circumstances consumed at the same time. As a consequence each purchaser is likely to receive a different service experience, or in other words the service is heterogeneous. Clearly the potential for such variability will hinder the process of evaluation by consumers (Ennew and McKechnie, 1998).

Internet banking, as explained earlier in section (3.6) under customisation, however arguably offers standard services, perhaps programmed for different groups of buyer, where performance will be near identical from one day to another, with perhaps the only observable change being the performance (i.e. the quality of connection) of the Internet itself. It was also argued earlier, that as a consequence of this apparent near homogeneous nature of Internet banking services, some Internet banks have to are attempting to customise the product by emphasising more on objective dimensions, such as interest rates, charges, etc. a favourable implication of this development is that the evaluation of Internet banking services can be based on more objective criteria.

4.5.5 Fiduciary Responsibility

The concept of fiduciary responsibility refers to the implicit and explicit responsibility of financial services organisations for the management of their customers' funds and the nature of the financial advice supplied to their customers (Ennew and McKechnie, 1998; McKechnie, 1992). In financial service transactions, a set of promises is essentially being exchanged between the buyer and the seller. From the buyer's point of view much depends on what exactly is being promised and the likelihood of such promises being delivered (Lewis, *et al.*, 1994). In the case of long-term savings plans for example, it is difficult for consumers to evaluate promises that are given in the absence of full information (McKechnie, 1992). Decisions on whether to purchase such services are more likely to be based on experience and credence qualities, as there are fewer search qualities. Before any financial resources change hands consumers must have confidence and trust not only in the financial institution concerned but also in its personnel. Apart from relying more on information from personal sources, consumers are likely to consider factors such as the size, longevity and image on the financial services organisation as indicators of whether any promises made are sound and are likely to be fulfilled. The establishment of trust can also bring about a degree of inertia in buyer-seller relationships. Since an irreversible amount of time and effort is required by an individual in order to acquire the necessary experience and information on which to assess an institution's reliability, it is usually the case that once satisfied, a consumer is more likely to remain with that institution than incur the costs of searching for and vetting alternative suppliers.

The impact of fiduciary responsibility is arguably at its greatest at the purchase stage (Ennew and McKechnie, 1998). Consider the situation, whereby a financial institution carries out an active marketing campaign, which has stimulated a decision by a consumer to purchase. Then the consumer may find that, despite the institution indicates that it is unable to provide the product. Partly this may simply be explained by the overall importance of profit and an unwillingness to supply a product (such as a loan, insurance, etc.), which the risk is too high (Knights, *et al.*, 1994). Ennew and McKechnie (1998) note that it is worthwhile to appreciate that such decisions may also reflect an element of fiduciary responsibility in the sense that financial services suppliers are obliged to recognise that many of their 'raw materials' are actually funds provided by other customers.

Furthermore there are difficulties for financial service organisations in that fiduciary responsibility means that that they may be promoting products to those individuals who are unlikely to be able to purchase them because they are considered to be poor risk. While clearly this is something that many suppliers seek to avoid, in practise the identification of exactly who is an appropriate customer is difficult and even with sophisticated marketing information systems, this process will be less than perfect.

The selling process itself is also an area of concern because of the substantial information asymmetries which exist between supplier and customer. To address these problems is difficult. Given the information asymmetries which exist between supplier and consumer, many consumers are vulnerable to high-pressure selling and bad advice. Indeed this is probably the issue that has done most to undermine the image of the financial services sector in recent years (Ennew and McKechnie, 1998). Nevertheless there are ways in which these issues can be tacked both internally and externally. The simplest route is perhaps to emphasise honesty and prudence as themes in promotional campaigns.

In Internet banking there appears to be blurred boundary between the responsibilities of the Internet banking service provider and the consumer. This is perhaps because Internet banking services are in their formative stages, hence both the employees of the banks and the customers are all going through a process of understanding the workings of the processes. For example, consider a simple case of funds transfer from one account held with the Internet banking service provider to another domestic bank. It is relatively easy on the part of the service provider to promise that funds transfer process will take only a few seconds to execute and that funds should reach the destination with a specified number of days. Suppose the funds do not reach the destination. The customer sends an email, expecting an instantaneous reply. The reply could, arrive late, not address the problem adequately, or in the worst-case scenario never receive a reply at all. Then the customer might attempt to use the telephone help line. The operators of these telephone help lines are not typical bank employees, and hence they may lack the adequate skill to explain the process and where things could have gone wrong or present a solution. An Internet banking service provider who has made a customer go through such an experience has failed in a number of aspects. They have not communicated what exactly is being promised and the likelihood of such promises being delivered, nor have they instilled customer confidence and trust not only in the financial institution but also in its personnel. Therefore, even in Internet banking services customers may be compelled to rely more on information from personal sources, consider factors such as the size, longevity and image of the organisation as indicators of whether any promises made are sound and are likely to be fulfilled.

Problems associated with the selling process take a different form in Internet banking services. The Internet is a pull medium, i.e. consumers go in search of information, unlike traditional media where the information is pushed towards the consumer. Hence the likelihood of high pressure selling is low, with the only persuasive form of selling being personally addressed emails. By emphasising honesty and prudence as themes in promotional campaigns, these problems may be minimised.

4.5.6 Two Way Information Flows

As far as two way information flows are concerned, what is unique about financial services in general is that rather than being concerned with one-off purchases, they involve a series of regular transactions between a buyer and seller, usually over an extended period of time (McKechnie, 1992). As a by-product of the normal operation of these transactions, a great deal of up-to-date private and confidential customer information is captured, which can subsequently be used to maintain and develop relationships with existing customers as well as attracting new ones. This position is enhanced in an Internet banking environment: not only is information more available, but also a greater amount of information can be collected, such information tends to be more accurate and up to date largely due to the automated nature. Further, the collection of the information does not involve additional expenses.

4.5.7 Long Term Nature

Many financial services are either consumed continuously (for e.g. current accounts) and therefore require a long-term relationship or else only yield benefits in the longer term. As indicated earlier, these features of financial services will tend to increase the perceived risk associated with the purchase and decrease the consumer's ability to evaluate the service both ex ante and ex post. To address this problem, there is again a tendency to rely heavily on marketing activities which emphasise the longevity of the supplier, trust, confidence and reliability.

This analysis has shown that while Internet banking services remain largely intangible as with traditional banking services, other traditional banking service qualities of inseparability, perishability and heterogeneity are only partially reflected in Internet banking services. Internet banking services tend to be more standardised to a considerable degree, with a clear separation between the provider and customer. Additionally, the consumer involvement and their input in the service delivery is very high. Hence any heterogeneity of a particular Internet banking service can be attributable to the consumer, the Internet service provider and any hardware used in the process of receiving banking services. It was argued that Internet banking services appear to have been 'stored' hence is not perishable. In light of the 'changed qualities', it could be argued that Internet banking services qualities are more akin to those of physical goods. Hence on balance, it can be said that Internet banking consumers enjoy benefits that can be evaluated objectively to a large extent, unlike traditional financial services that are usually evaluated using subjective criteria, such as attitudinal research. This is an important argument that is developed further in the next chapter.

4.6 Chapter Summary

This chapter achieves two objectives. Firstly, it examines problems that consumers face as buyers of financial services. Secondly, it examines the qualities of Internet banking services.

There have been a variety of empirical research studies examining consumer behaviour and choice in financial services. It is apparent from these studies that, both conceptually and empirically, there are certain qualities of traditional financial services that present a number of problems for consumers when they make choices. It is against this backdrop that the qualities of Internet banking services are explored.

It was argued that Internet banking services portray qualities that are more akin to physical goods. They contain comparatively higher levels of search qualities in comparison to traditional financial services. Further, attributes like experience qualities are relatively easily acquired in Internet banking services, because the demonstration component enables a simulation of purchases and operation of accounts. Information collection is made easier, and the need to rely on the experience of others is significantly reduced.

Having examined the qualities of Internet banking services, the next chapter generates testable hypotheses to address the central research question.

Chapter 5.

The Research Question and Working Hypotheses

5.1 Introduction

The last chapter built upon the literature reviewed in chapter three and applied these concepts to financial services in general and Internet banking services in particular. It examined the Internet environment and emphasised the qualities of Internet banking services by relating to conceptual foundations and to empirical research work in consumer behaviour in the financial services sector. The objective of this chapter is to draw out gaps, highlight areas where more work and a better understanding is required. Each of these gaps or areas gives rise to hypotheses that encapsulate the central research question that requires attention. The hypotheses are necessary to develop a framework within which the methodology to be constructed, which becomes the central contribution of this thesis.

This chapter is divided into three sections. The first section examines the need to approach the development of the methodology framework by generating hypotheses. In order to generate hypotheses, the research question needs to be revisited. Accordingly, the second section summarises the main findings of the literature and thereby highlights the importance of the research question. Thereafter the third section begins to develop hypotheses. For this purpose, the third section is subdivided into three parts. The first part revisits consumer behaviour models and establishes the decision-making process that a cognitive consumer goes through in the Internet banking services purchase process. This is an important exercise in highlighting the background to the research question. The next two parts are important in exploring gaps that give rise to testable hypotheses to address the central research question. In the second part, the information collection stage and its importance is examined, and in particular how this process takes place in an Internet environment is discussed. And finally, once the information has been collected, an evaluation takes place before the purchase. In comparison to the terrestrial world, the crucial difference here is that some of the post purchase evaluation aspects, such as service quality of the service can even be assessed prior to purchase. Each of these parts continue to build on the literature review carried out in previous chapters.

The chapter concludes with a summary of the findings of the chapter.

5.2 Hypotheses and the Development of a Methodology Framework

It is good research practise to draw up hypotheses based on the literature review, and thereafter to test such hypotheses using a suitable research methodology. The results of such tests form the findings and conclusions of most pieces of research. This thesis differs from this formula by a considerable degree.

In this thesis, the literature review has already been carried out in chapters 2, 3 and 4. It is now logical to develop the hypotheses by identifying gaps in the literature. However, as it has already been stated in the introductory chapter, due to the characteristics of data, full hypotheses testing cannot be undertaken. Therefore, the most important contribution of this thesis will be the development of a methodology framework to examine Internet services in general and Internet banking services in particular. It therefore follows that, while the hypotheses are important, the primary use of the hypotheses are to develop a framework within which the methodology can be constructed. As alluded to earlier, the testing of hypotheses is of secondary importance, primarily because the results of such analysis can only be of indicative nature at this point in time.

The rest of the chapter is devoted to the development of hypotheses. A question that may arise at this juncture is: why is such an importance paid to the development of the hypotheses, if they are not an end, but in large part a means to develop the methodology framework? This can be defended on two grounds. Firstly, an extensive literature review was carried out, and therefore a considerable number of gaps in knowledge that merit further investigation has been identified. And therefore to maintain academic rigour of this research these gaps must investigated by means of hypotheses. Secondly, due to the nature of the proposed research methodology (this will be discussed in detail in the next chapter) the quality of the methodology framework in depends on large part on the extensiveness of the questions being investigated. And therefore the more attention is paid to the development of hypotheses the more refined the methodology framework would tum out to be.

The next section of the chapter lays the foundation to the development of the hypotheses by re-visiting the research question.

5.3 The Research Question

The literature review in previous chapters has explored the characteristics of the banking sector uncovering threats and opportunities presented by the changing environment in which banks operate. It was seen that the combined forces of technology, information, changing markets and regulation, are eroding many of the traditional reasons for financial intermediation. In particular, the analysis highlighted the changing nature of technology in banking, the increased importance and impact that new Internet bank entrants are having on the banking sector.

This discussion on the banking sector was extended in section (4.5), where qualities of Internet banking services were examined. It was found that unlike traditional financial services Internet banking services portray qualities that are more akin to physical products. They contain comparatively higher levels of search qualities in comparison to traditional financial services. Further, attributes like experience qualities are relatively easily acquired in Internet banking services, because the demonstration component enables a simulation of purchases and operation of accounts. Information collection is made easier, and the need to rely on the experience of others is significantly reduced.

The literature review in chapter three demonstrated that there is no single explanation of consumer behaviour leading to a purchase. Additionally, it was also seen that holistic approaches to the study of consumer behaviour, the determination of customer choice and purchase has been relatively under-researched (Gabbot and Hogg, 1998). Therefore, it became evident that a structured approach to the analysis of understanding consumer purchase intentions was desirable, and an information-based perspective, the cognitive consumer perspective was chosen. The information-processing approach offers a clear framework to structure the discussion. The choice was defended on the grounds that in addition to being accessible it is both well documented and is the dominant paradigm. An analysis of consumer behaviour in traditional financial services was carried out in section (4.2). It was found that both conceptually and empirically, there are certain qualities of traditional financial services that present a number of problems for consumers when they make choices.

More succinctly, the main conclusions that can be drawn from the review are as follows.

1. There are many changes taking place in the banking sector, and as one of the responses to these changes, banks are increasing their reliance on technology, more specifically Internet technology. These developments have paved the way to Internet banking services.

2. While Internet banking services share some of the traditional financial service qualities such as intangibility and fiduciary responsibility, they portray some qualities that are more akin to physical goods.
3. There are many competing theories attempting to explain consumer behaviour leading to a purchase. The concept of the cognitive consumer was argued to be a useful framework for analysis.
4. Empirical research in consumer purchase of traditional financial services emphasises that consumers face many problems when making their choices.

It was also argued in section (4.3) that empirical research examining consumer purchases in financial services tended to examine a single aspect of a service. There appears to be a clear void in empirical research that takes a holistic approach to purchase decisions in financial services. Perhaps, this is due to an aspect discussed above, most notably that consumers face many problems when purchasing financial services, and therefore carrying out empirical research is difficult. As argued above, Internet banking service characteristics are significantly different from those of traditional financial services. Therefore, these observations give rise to the central research question: **what characteristics of an Internet banking service influence consumer purchase intentions?**

In order to address this research question, this thesis has so far laid the foundation by reviewing literature on a number of interconnected areas of study. The following sections brings together this body of knowledge, identify gaps, and develops testable hypotheses. Collectively, the process of testing these hypotheses provide a solution. to the central research question. More importantly, the hypotheses provide a framework within which the methodology can be developed.

5.4 Consumer Purchase Behaviour - A Revisit

The concept of the cognitive consumer is based upon a fundamental assertion that the purchase process is a sequence of tasks characterised as a problem-solving exercise. The purchase problem is formulated as a desired outcome where a consumer must evaluate the components of the environment in order to construct a solution or course of action to achieve the desired outcome and comprises of a number of distinct stages. Essentially there are four stages to this framework; needs recognition, identification and evaluation of the alternatives that can fulfil the needs, purchase of a service that can satisfy the needs, and finally post-purchase evaluation.

The first stage is the recognition of a need that can be fulfilled through the consumption of an Internet banking service product. Since the central research question attempts to determine what features of an Internet banking service influence consumer purchase intentions, it is taken as given that the consumer has already decided to purchase an Internet banking service product. The need to consume an Internet banking service product may be in response to a specific need or a collection of needs: for example, such as the need to have an account to carry out services such as bill payments, transfer of money, withdrawal and deposit of money, etc.

The second stage commences with the consumer collecting information about alternatives from various sources, followed by an evaluation stage where the consumer engages in some sort of evaluation process to arrive at a ranking of the identified product alternatives. The primary source of information in purchasing an Internet banking services is the Internet. Information on the Internet can take many forms, including, the bank's home page, information search through search engines and portals, newspaper like articles found in regular online publications, etc. Additionally, information can also be collected from "traditional" sources such as advertisements in traditional forms of media, direct marketing, other people, trade magazines, company databases, etc. Following the example above, at this stage the consumer will come to the conclusion that only a current account will fulfil their needs. The importance of information collection is highlighted in the evaluation stage. For the comparison of alternatives to take place, it is important that information relating to all aspects of the Internet banking service is collected. Internet banking service providers will only be successful in selling their products if they are able to communicate product attributes through all possible means. Similarly, the consumer is assumed to be using a wide range of sources and types of information. This process of evaluation continues right up to the purchase of the product.

The pre-purchase stage of consumption concludes with the purchase itself It was observed that some authors contest the inclusion of this behavioural activity as a separate stage on the grounds that it is difficult to extract it from the previous stage. Unlike traditional service products which are low in search qualities, Internet banking service products are high in search qualities hence the distinction is more valid. In addition, it focuses attention on the acquisition of the service products, the payment method or the exchange mechanism, delivery, the Web interface and role of staff and other people manning the back office. Further attention is also on the time taken and the interaction that takes place. Although this stage is not the emphasis of this research, it deserves attention for reasons outlined above.

Finally, once the purchase is complete, the post-purchase assessment takes place.

This simplified purchase process allows complex purchase behaviour to be broken down and paves the way for a meaningful analysis. Such an analysis also assumes that individuals are fully aware of their actions and act in a rational manner. Such an assumption is valid due to the nature of the service procurement process and the more active role that a consumer will have to adopt to complete the purchase process. For instance, it is extremely unlikely that a consumer will purchase an Internet banking service product purely on an impulse, while it is possible such a consumer may purchase a product from a terrestrial shop after observing it through a shop window. Additionally, the decision format identified above is predicted on the assumption that products are inherently similar in terms of the decision process. This assumption is supported on the grounds that; firstly, the consumer is seeking to purchase a single type of product; secondly, the products are presented in a largely similar environment, i.e. over the Internet; thirdly, as described in the previous chapter some of the qualities Internet banking service products are akin to those of physical goods.

The following sections of this chapter examines each of these stages in tum with a view of formulating hypotheses to achieve twin objectives. Firstly, and most importantly, the hypotheses act as a framework for developing the research methodology, and secondly, the hypotheses are used to tentatively examine the central research question.

5.4.1 Information Search and Acquisition

Faced with any purchase decision, a consumer requires information with which to inform product choice. Consumers may acquire information about alternatives from various sources. The primary source of information in this context is the Internet, and it will be the focus of this section. The primary source of information on the Internet is the bank's home page and the hyperlinks associated with it. Information search can also be carried out through the use of search engines and portals, newspaper like articles found in regular online publications, etc. However, the process of information collection from sources other than the primary source, the bank's home page, is fraught with danger for two reasons. Firstly, because the Internet is a vast repository of information, there is a risk that the consumer may be overloaded with information. Secondly, there are very few means of ascertaining the authenticity and credibility of such information. Therefore a customer collecting information should be mindful of these pitfalls. As stated earlier in section (3.6) and reiterated above, the importance of information collection is highlighted in the evaluation stage. Comparison of alternatives can take place *only* if information relating to all aspects of the Internet banking service is collected.

The analysis in section (4.5) illustrated that Internet banking services are comparatively high in search qualities compared to traditional financial services. Search qualities are attributes that can be determined prior to purchase. They are usually associated with factual information that is common to all consumers and verifiable in advance of purchase. It is important to note that information on Internet banking services, unlike in traditional services do directly describe the experience that the consumer will actually receive to a very large extent.

Information search can be enhanced by relating to *experience qualities,* which are attributes that can traditionally only be determined after purchase or during consumption. However, as stated on elsewhere in sections (3.5), (3.6) and (4.5), experience in Internet banking services can be gained through a pre-purchase trial. Although a pre-purchase trial is not a common option in the case of traditional services, it is possible to simulate a service episode in Internet banking services to establish whether the consumer will purchase a full service or a sequence of episodes. This process is facilitated by a demonstration component in the Internet bank's Web site. Additionally, similar to the free inspection of physical products where the trial is exactly equivalent to the actual product, in Internet banking services there is a very large likelihood that the rest of the service will deliver the experience of the trial. Unlike other services, where the participation of any other individual gives no guarantee of a repeated performance, in Internet banking services performance is largely homogeneous, hence observation is a reliable source of information. As a consequence, contrary to other services the need for a consumer investigation of the experiences of others is reduced.

Therefore, it is hypothesised that,

H1: Purchase intentions, mediated by the perceived information search and experience quality, are influenced by the contents and presentation of the demonstration component of an Internet banking service.

This process of information search leads the consumer to an evoked set of alternatives that are likely to fulfil their expectations that form the basis of comparison and choice. Aspects of comparability (Bitner, 1992; Lewis, 1991) of Internet banking services were discussed in section (3.5.2). Consequently, similar to the case with physical goods, Internet banking services are likely to result in a considerably large evoked set. A positive aspect of this quality for researchers is that it significantly reduces arguments to employ attitudinal research methodologies to test such constructs. However, a negative effect that accrues to banks is that consumers may prove to be less loyal once they find a superior alternative (Zeithaml, 1981).

The literature review in section (3.5) also examined various models of how consumers choose between available alternatives in different situations. The common denominator of these models is a set of attributes. However, there are problems for consumers in defining attribute sets in relation to services. To begin with there are problems of identifying what the attributes are in the first place. If this hurdle can be overcome, there are problems in making comparisons on the basis of these attributes.

Although attributes of Internet banking service products are to some extent still intangible, these attributes can be determined prior to purchase and are common to all consumers purchasing the product. Therefore, these portray characteristics of tangible attributes. On the other hand, it is also true that a consumer's needs are accommodated by their involvement in service delivery, therefore the attribute will also be a function of that individual's involvement. For example, when an Internet banking service customer instructs a bank to execute a domestic money transfer, the customer is involved in describing the recipient bank account, the date the customer want the transfer to take place, and the amount of the transfer, etc., and an erroneous instruction will be directly reflected the service delivery.

It was also suggested that consumers find it difficult to compare service alternatives on the basis of common attributes. This was because, it was argued, a consumer cannot put two services side by side at any one time, hence they cannot be compared simultaneously, but can only be compared in series, not parallel. As reiterated on several occasions, Internet banking services portray the qualities of a physical product, since it is possible to compare two Internet banking services side by side at any one time and they can be evaluated simultaneously. Further the presence of truly common attributes implies that Internet banking services are comparable products.

The above discussion demonstrates the importance of information search and acquisition in determining the purchase determinants in Internet banking services. This is one of the central tenants that drive this research. i.e. through a process of information search, collection and evaluation, consumers are able to determine their purchase intentions in Internet banking services.

Therefore, it is hypothesised that,

> ***H2: Purchase intentions, mediated by perceived information search quality, are influenced by the contents of an Internet banking service Web site.***

5.4.1.1 Technical Qualities of Banking Services

The literature review in section (4.3) demonstrated that for traditional financial services, the relative lack of empirical support for the importance of technical qualities (i.e. what is actually delivered or the functionality) of the financial services and their influence on purchase intentions. Consumers were seen to be more interested in service qualities or how the financial service is delivered. In other words they are more interested in the dominant quality of financial services, i.e. the personal qualities of financial services. A possible explanation for this lack of evidence to support for the importance of technical qualities may be as follows: traditional financial services are intangible, inseparable and heterogeneous, hence there are fewer tangible cues on which to base decisions and therefore greater reliance is placed on experience qualities after purchase or during consumption. Even in the case of financial services that are high in credence qualities, experience qualities may take prominence since consumers may find it extremely difficult to make retrospective assessments in hindsight, i.e. whether they entrusted the right financial institution with the right management with their financial resources.

However, consumers have the ability to assess the performance benefits of Internet banking service more objectively. Firstly, as described earlier, the performance of an Internet banking product is relatively easy to predict prior to purchase, because many Internet banking service Web sites incorporate demonstrations that illustrate how different transaction procedures can be carried out. This process gives the consumer the ability to compare the performance on an objective basis, compare and contrast different services from different Internet banking service providers, etc. Consequently, the problems with evaluating functional qualities are mitigated, and technical qualities including the functionality become more important. Secondly, consumers need enhanced functionality in their Internet banking services to cater for their diverse needs. Therefore, it is logical to conclude that increased technical qualities, increased functionality has a strong influence on purchase intentions. Thirdly, as described elsewhere, an Internet banking service is a largely mechanistic service, which is for all intents and purposes, delivers a homogeneous service. The consequences of this routinization of Internet banking services are that consumers are likely to be less interested in how the Internet banking services are delivered. Finally, it was argued in section (3.6.1) that the influence that Internet banking services have on the service context (Bitner, 1992; Zeithaml and Bitner, 1996) is limited. The consumer is in charge of the physical surroundings and the equipment used to consumer the service. The bank's can influence the service context by controlling the presentation of the functionality of accounts. Therefore, a method of differentiating between different Internet banking services is through an evaluation of its technical qualities.

Therefore, it is tentatively hypothesised that,

> *H3: Purchase intentions, mediated by perceived technical qualities, are influenced by the functionality of banking service products of an Internet banking service.*

5.4.1.2 *Pricing and Perceived Value in Internet Banking*

The theoretical interpretation of the effects of price on purchases, as discussed in section (3.3.1), postulates that consumer preferences for an Internet banking service over a conventional terrestrial banking service would be determined by consumers' tastes (and represented by economists' indifference curves) and by relative prices (budget line). Holding other factors constant, an innovation like Internet banking, which reduces the costs of service provision significantly, makes it significantly cheaper than terrestrial banking services, i.e. relative prices in the budget line tilt in favour of cheaper Internet banking. Ina new equilibrium for these close substitutes, purchases (assuming normality) would switch in favour of the Internet provision. If the price of Internet banking drops (for example higher interest rate for savers, free-of-charge current accounts, low interest on current account overdrafts), the number of accounts demanded by the consumers will increase. This means that own-price elasticity of Internet banking also has a strong influence on the quantity (number of accounts) demanded, and is given by the demand curve assumed to be downward sloping for normal goods. Just as price may have an enormous influence in the shift on consumers from conventional banking services to Internet banking services, it also dictates the shift towards the particular Internet banking service.

It was argued in section (3.3.1) that consumer purchase intentions may also be affected by a bandwagon effect. Consumers may purchase from a particular Internet banking service when they observe that others are buying it. In this context the quantity sold (i.e. number of customers) by a particular Internet banking service provider communicated through various forms of media is a powerful tool for attracting new consumers to that particular bank.

Price is also closely related to value. The importance of conveying the perception of value in relation to bank's prices was discussed. The complexity of the Internet banking service offering, perceived knowledge and sophistication of consumers are crucial elements in perceiving the value of an Internet banking service. For instance an overly-complicated Internet banking service, makes it potentially difficult to envisage the use of specific service features as the basis for establishing a value, due to lack of consumer cognition.

Consumer satisfaction was recognised by Kotler and Levy (1969) to be highly associated with "value", and hence on "price" while on the other hand service quality is not generally considered to be dependent on price (Anderson *et al.,* 1994). Fornell (1991) and Lewis (1991) propose the element of price in the information of service quality.

Therefore, it is hypothesised that,

H 4: Purchase intentions are influenced by the prices (as measured by interest rates, fees, etc.) of an Internet banking service.

Here, higher savings account interest rates, higher interest rates positive current account balance and lower interest rates for overdrafts are equivalent to lower prices and is hypothesised to positively influence purchase intentions.

5.4.1.3 Brands and Risk in Internet Banking

Brands are becoming increasingly important as a means of attracting potential consumers (Palmer, 1988). As discussed in section (3.3.5), previous studies have shown that established brands have a positive effect on quality perceptions, and advertising was found to create an awareness of the brand. For instance, a heavily promoted Internet banking service brand increases the probability that the brand is included in the consumer's evoked set. A number of researchers, *Light (1990),* Nelson (1974), Kirmani and Wright (1989) found that advertising influences consumer perceptions of the quality of brand, and influence usage experience. In the goods sector, there is a long history of research on the relationship between advertising and sales, although this has not been replicated in the services sector.

Many business entities operating m the Internet environment appear to spend large amounts of money on advertising and other marketing campaigns through online and offline media and through other sponsorships. The rationale behind this investment for the future is based on the notion that it is extremely important to build awareness among consumers and "sign-up" as many customers as possible. Such a strategy assumes that the first mover advantage will propel these organisations into a commanding market share in years to come. Further the anticipation is that all these customers will generate a substantial stream of revenue in future. The discounted present value of projected revenue from these customers has been translated into extremely optimistic market valuations for these businesses, although these companies currently do not make any profits. Internet banks appear to be no exception to this rationale, a significant number of banks have carried out high profile advertising campaigns on the Internet. The objective of such campaigns is to raise the awareness of brand and the banking service products on offer, in the hope that they would influence consumer purchase intentions. As of early 2000, there were no audited advertising spending amounts by Internet banking services. Therefore, it is not possible to evaluate the direct influence of this variable on consumer purchase intentions.

Search engines are the most important search tools in the Internet used for information search and collection. If an Internet banking service brand is included in a consumer's evoked set, as a result of measures outlined above, but fail to be captured by a search engine, a consumer may be unable to locate the particular Web site. By extension such an outcome may negatively influence purchase intentions.

Hence, it is hypothesised that,

H 5: Purchase intentions are influenced by the visibility (as measured by capture by popular search engines and linkages to other Web sites) of a brand of an Internet banking service.

When consumers deal with banking services, and Internet banking services in particular, certain risk perceptions arise. As discussed earlier in section (3.3.2), these risk perceptions can range from the potential for their deposit monies to be lost to the likelihood of recurrent annoying technical errors. Such manifestations may be at their highest when dealing with a new Internet bank. This aspect is re-iterated by Devlin, *et al*, (1995), they argue that in the case of financial services the institution brand is more important than specific service product brand, hence the close association with the institution. Erdem (1998) contends that, therefore, a multi-product bank can use its *umbrella* brand as a bond for quality when it introduces a new Internet banking service. In this case, consumers use their experience with the parent institution service products as a signal of the quality of its extension. While some Internet banking services may use their parentage to reduce risk perceptions, increase exposure, instant recognition, others make very little effort to even hint at their parentage.

Nevertheless, it is hypothesised that,

H 6: Purchase intentions are influenced by the association (or proximity) of an old established umbrella brand.

Experience with the parent product thus provides consumers with information about the new product, this might reduce consumer uncertainty and, thus perceived risk associated with the extension (Erdem, 1998; Laroche *et al.*, 1996). As the brand becomes established, consumer risk perceptions may reduce. Besides any umbrella effect from a proximate brand may enhance the strength of the Internet banking service.

A new Internet banking brand may be supported by three phenomena. Firstly, a new brand may appeal to a completely new audience of consumers, who may be dissatisfied with traditional brands, leading to a favourable reaction from consumers. Secondly, a *bandwagon* effect may prompt more consumers buying new banking services from the Internet bank when they observe others are buying it. Thirdly, with a *snob effect,* a consumer to may want to appear different from the traditional bank consumer. This creates a positive impact on Internet banking services. On the other hand, the strength of an Internet banking brand may be positively related to how long brand has been in existence. The rationale here is that the longer the brand has been in existence, the longer the exposure, and the more confident people would have become with its existence and operations. Hence initially, a 'new' brand may be handicapped, but with the passage of time more consumers will begin to join the Internet banking service.

On balance, it is hypothesised that,

> ***H 7: Purchase intentions are influenced by the strength of a brand, as proxied by the time elapsed since brand launch.***

5.4.2 Evaluation of an Internet Banking Service

Once the information has been searched, acquired and evaluated the service can be purchased. Although post-purchase evaluation is not the focus of this thesis, the unique qualities of Internet banking services, as explained in section (4.5), enable the evaluation of some of the post purchase service qualities (for example, service quality) to take place prior to purchase. This is made possible by the demonstration component, which enables the consumer to experience account operation and purchase procedures through simulation.

In an Internet banking service encounter, personal interactions take place without the need for physical proximity. Although this form of encounter is not as rich as a face-to-face encounter, facilities such as email, Web messages, etc. can be utilised to significantly increase in the quality of communication.

There is a large degree of customer participation with regard to service delivery in Internet banking services. Traditionally, the customer interacted with bank employees in a particular service context. This shift in the level of participation has undoubtedly led to the nature of the service itself, since many researchers have concluded that one of the most important characteristics of the service product is the role of the customer in its delivery (Parasuraman, *et al.*, 1985; Bitner, *et al.*, 1990; Gronroos, 1978). Indeed, research by Gabbot (1996), Zeithaml and Gilly (1987), Hiltz and Johnson (1990), suggest that participation is as much a consumer variable as it is a product attribute.

This construct, the consumer engagement with a product, be it cognitive, emotional or behavioural, is widely used in explaining different degrees of activity in relation to product choice and consumption (Gabbot and Hogg, 1998). The level of engagement will vary with the type of service. Apart from the type of service, factors such as mood, personality and individual resources will also play a part in determining the degree of interest shown by the consumer on a particular service. It was seen above that there can be several levels of proximity in a service encounter, but every service encounter will take place within a service context, and at a certain level of engagement, which then becomes an integral part of the service encounter experience, whether or not it is intended to be so. These aspects are best evaluated by examining the service quality of Internet banking services, and hence the next section examines the influence of service quality of Internet banking services on purchase intentions.

5.4.2.1 Service Quality in Internet Banking

Section (3.6.3) discussed the importance service quality in services. Section (4.2) contributed to this discussion, by investigating the considerable ·number of empirical studies examining service quality in financial services. These analyses concluded that, (1) current performance adequately captures consumers' perceptions of service quality offered by a specific service provider, and (2) service quality is an antecedent of consumer satisfaction, which in turn exert a strong influence on purchase intentions, (3) in addition to the theoretical support for the performance-based measures of service quality, practitioners often measure the determinants of overall satisfaction/perceived quality by having consumers simply assess the performance of the company's business process. This means that service quality of an Internet banking service may be evaluated by measuring the performance of an Internet banking service account.

Empirical studies emphasised the importance of the five dimensions (initially presented by Parasuraman, *et al.*, (1988)) of service quality, i.e. tangibles, reliability, responsiveness, assurance and empathy, and their importance in measuring service quality. In a less referenced and a more recent study, Avkiran (1999), presented four dimensions, staff conduct, credibility, communication and access to teller services, to measure service quality in banking services.

When banking services are delivered over the Internet, these service quality dimensions take a different configuration. While the tangible dimensions of a terrestrial banking service are the physical facilities, equipment and appearance of bank personnel (Ennew and McKechnie, 1998; Shostack, 1982), in Internet banking, the equivalent tangible dimensions are on-screen Website attributes, and these attributes replace their terrestrial counterparts.

Therefore, it is hypothesised that,

H 8: Purchase intentions, mediated by perceived Internet service quality, are influenced by Website characteristics.

Reliability and responsiveness in terrestrial banking refers to the ability to perform the proposed service dependably and accurately, and the willingness of bank staff to help bank consumers and provide prompt service, respectively (Czepiel, *et al.*, 1985; Hochschild, 1983). Avkiran (1999) defined staff conduct as, responsiveness, civilised conduct and the presentation of branch staff that will project a professional image to the consumers. Equivalent Internet banking dimensions that would correspond to these attributes are functions of Web pages, hypertext attributes, and a collection of software and hardware that supports the entire operation.

Assurance in terrestrial banking refers to the knowledge and courtesy of employees and their ability to inspire trust and confidence. Empathy pertains to the caring, individualised attention that a bank may provide for its consumers (Hochschild, 1983; Ashforth and Humphrey, 1993). Avkiran (1999) defines credibility to signify the maintenance of staff consumer trust by rectifying mistakes, keeping consumers informed, and communication as the process of fulfilling the banking needs of consumers by successfully communicating financial advice and serving timely notices. These dimensions are quite difficult to offer through an Internet medium. The entire Internet banking operation should demonstrate assurance and empathy through the provision of 'consumer friendly' opportunities for interaction and communication.

Therefore, it is hypothesised that,

H 9: Purchase intentions, mediated by perceived Internet service quality, are influenced by the performance of the Website and related components.

Further, that,

H 10: Purchase intentions, mediated by perceived Internet service quality, are influenced by the responsiveness of communication, speed and accuracy of responses to customers.

Privacy and Security

A crucial element that is essential to the formation and maintenance of a banking relationship between a consumer and a bank is security. An Internet banking service must be able to instil confidence in consumers by adequately informing the consumer of the measures in place to protect both confidential information and most importantly their assets. For instance, when communicating with the bank the customers may use email or a Web based application. If they require to communicate with confidential information measures should be in place so that the communication application is secured with adequate means, along with assurances that privacy is maintained. The key element is to demonstrate to the consumer that they are being individually treated in an honest, speedy, secure and accurate manner. Additionally, attributes like brand name (discussed above) assist in instilling confidence in consumers.

Therefore, it is hypothesised that,

H 11: Purchase intentions, mediated by perceived Internet service quality, are influenced by Website security.

5.5 Chapter Summary

This chapter is crucial to this thesis in formulating working hypotheses, which are crucial in formulating a framework for the development of a research methodology. This process was achieved by, firstly, highlighting gaps in the literature review in previous chapters and, secondly, by emphasising areas where a better understanding is required by building on the present knowledge. Each of these gaps or areas gave rise to hypotheses that encapsulated the central research question. In this manner eleven hypotheses are developed to test for purchase intentions. It hypothesises that purchase intentions are influenced by a combination of information search and experience qualities, technical qualities, pricing, branding and service quality of an Internet banking service. Taken together these hypotheses underpin the central research question.

Evaluation of Internet banking services is complex, but nevertheless for reasons outlined earlier will offers exciting opportunities in terms of research methodology. This is the subject of analysis in the next chapter.

Chapter 6.

Towards a Research Methodology

6.1 Introduction

The previous chapter highlighted gaps in the literature and emphasised areas where more work and a better understanding is required. Each of these gaps or areas gave rise to hypotheses that encapsulated the central research question. The objective of this chapter is to develop a research methodology that can test these hypotheses.

This chapter is divided into five sections. The first section illustrates the background of the methodology problem, by highlighting the need to investigate all possible options in light of the significant differences in the qualities of Internet banking services compared to traditional banking services. It is customary to approach such problems by looking at previous work. Accordingly, the second section examines previous empirical research work in traditional banking services. Particular attention is paid to data collection methods, data analysis methods and how methodological justifications were made. The third section examines the case for adopting one or a combination of reviewed research methodologies to Internet banking services. In doing so it presents arguments for the need for an alternative research methodology to address the central research question of this thesis, in light on the unique qualities of Internet banking services and the opportunities it presents. Section four introduces, explains and charts the development of 'mystery-shopping', the term given to participant observation in the commercial sector, and adopted for this research as the preferred data collection methodology. This section also illustrates the growing use of mystery-shopping in the traditional banking services sector. The final section advance the discussion further by examining the suitability of this method for the present research question by; (1) adapting the mystery-shopping research methodology for Internet banking services, (2) discussing strengths and weaknesses of the methodology, and (3) how the research measures up to the principles of good research design. The chapter concludes by summarising the key observations.

6.2 Research Methodology: The Background

As stated above, the objective of this chapter is to propose a research methodology to test the hypotheses generated in the previous chapter. The proposed methodology will have to take into account the qualities of Internet banking services (discussed in section (4.5)) and the dimensions of an Internet banking service encounter (discussed in section (3.6)). Therefore it is useful to briefly revisit these issues and appraise implications that these may have on data collection, and by extension on the research methodology.

1. Internet banking services are intangible and a customer consumes the service while they are on the Web site. These Web sites are subject to frequent modifications, on occasions without leaving a trace of preceding offerings. Therefore the data collection methodology proposed must take into account the dynamic nature of the Web sites.

2. It was argued that Internet banking services contain significantly higher levels of search qualities in comparison to traditional financial services. Additionally, attributes such as experience qualities are relatively easily acquired in Internet banking services. This makes the data collection process much easier and more objective.

3. Related to above, it was also seen that it is possible to compare and contrast different Internet banking services by evaluating them side-by-side. This makes the evaluation of the service more akin to the evaluation of a physical product. Therefore more objective measurement criteria can be incorporated in the data collection instruments and consequently may reduce emphasis on attitudinal research.

4. Internet banking has been in existence only for a short time. An implication is that time-series observations are unavailable, for they would not yield sufficient number of observations. Therefore, meaningful time series studies cannot be undertaken until the industry matures.

5. It was seen that, unlike in traditional services, the service context in an Internet banking service is beyond the control of the bank to a considerable degree. Each service context unique. Therefore, it is difficult to maintain an uniform service context during the data collection stage that may be required for generalisable results.

6. Empirical studies that evaluated service quality in traditional banking services employed attitudinal research methodologies. However as the hypotheses show, service quality in Internet banking services

involve the measurement of performance attributes. The performance of these attributes can be measured using objective criteria.

While the above is by no means an exhaustive list of how the nature of Internet banking services differs from traditional banking services, it is clear that there will be implications if research methodologies in traditional banking services are directly applied to Internet banking services. When faced with such circumstances, it is prudent to investigate research methodologies adopted to examine similar research questions in traditional banking services. Once the investigation has been carried out it will form the base that will aid in the determination of an appropriate research methodology to test the hypotheses in this research. Therefore, each previous research methodology is examined in detail with particular attention to data collection methods, data analysis methods and how researchers made methodological justifications.

6.3 Methodologies in Traditional Banking Services

Traditional banking services have portrayed all the qualities of a generic service, i.e. intangibility, inseparability, heterogeneity, perishability and the concept of ownership (Sasser *et al.*, 1978; Lovelock, 1981 and Gronroos, 1978). Additionally, banking services were seen to be high in credence qualities and an element of fiduciary responsibilities (McKechnie, 1992). Hence, in the opinion of a number of researchers, these qualities have meant that in order to address issues such as service quality and consumer satisfaction in banking services, large-scale attitudinal surveys needed to be undertaken. The majority of previous research has relied on (1) structured questionnaires, administered either directly by the respondent, or by an interviewer (2) structured, semi-structured interviews and focus groups, and (3) case studies of institutions where initiatives have been carried out with specific aims.

The selected research for methodological discussion in this chapter has been drawn so that it meets the following criteria. Firstly, they comprise the more recent studies. Secondly, as stated earlier, these research studies have looked primarily at the banking services sector. Finally, where the research has addressed the services sector as a whole, the banking sector has been one of the industries within the larger sample. Additionally these studies are considered to be very influential in understanding the fundamentals of services marketing by the academic and commercial world.

6.3.1 Data Collection Methodologies

As noted above, empirical research in traditional banking services appears to have relied on three primary methods of data collection: questionnaire surveys, interviews and case studies. The following three sub sections will examine these three methodologies, taking each methodology in tum by grouping similar studies.

6.3.1.1 *Questionnaire Surveys*

A number of studies have utilised questionnaires as an instrument to measure attributes such as service quality and consumer satisfaction, and the effect of these on purchase intentions in banking services. Some of these studies are listed in Table (6.1).

Title of Study	Author(s)
Developing an Instrument to Measure Consumer Service Quality in Branch Banking	Avkiran, N. K. (1994)
Quality Consumer Service Demands Human Contact	Avkiran, N. K. (1999)
Investigating Drivers of Bank Loyalty: The Complex Relationship Between Image, Service Quality and Satisfaction	Bloemer, J., Ruyter, K.de and Peeters, P. (1998)
How Consumers Choose Banks	Khazeth, K. and Decker, W. H. (1992)
Determinants of Consumer Satisfaction in Retail Banking	Levesque, T. and McDougall, G.H.G. (1996)
Service Quality: Student's Assessment of Banks and Building Societies	Lewis, B.R., Orledge, J. and Mitchell, V.W. (1994)
The Behavioural Consequences of PC Banking	Mols, N.P (1998b)
SERVQUAL: A Multiple-Item Scale for Measuring Consumer Perceptions of Service Quality	Parasuraman, A., Zeithaml, V.A. and Berry, L. (1988)

Table (6.1): Studies utilising questionnaire surveys as the instrument of measurement

Most of these studies adopted a two-stage methodology. Firstly, the questionnaire instrument is pre-tested or tested as a pilot instrument. Depending on the feedback, appropriate alterations are made. This serves as a useful method to increase the clarity of the instrument before being fully deployed in a full-scale survey. During the pilot stage, sample sizes tended to be generally small, for instance, Mols, (1998b) made the pilot study among 30 respondents. However Parasuraman, *et al.,* (1988) had a sample of 200. The Parasuraman *et al.,* (1988) study was unique, since it used the pilot study to refine the items in the questionnaire through statistical analysis, before proceeding on to a second stage of the research, with a refined questionnaire item list, involving a further 200 subjects. The largest sample for a full survey is 2500 by Bloemer, *et al.,* (1998).

Among the different research studies, the number of items on the questionnaire shows a wide variation. Parasuraman, *et al.,* (1988) in their exploratory studies had 97 items, while Levesque and McDougall (1996) had only 17. Some researchers, for example Khazeth and Decker, (1992) divided questionnaires into different sections, each section concentrating on a specific research question. Almost all researchers collected demographic information on the respondents. Immediately after the publication of the seminal SERVQUAL study by Parasuraman *et al.,* (1988), there was an explosion in the number of publications examining service quality in all service sectors, and banking was no exception. Many of the banking sector studies retained the generic items proposed in the SERVQUAL scale in an attempt to replicate the original findings. In all cases, these studies found that, although Parasuraman *et al.,* (1988) formulated the SERVQUAL scale as a generic service quality measurement instrument, item changes were needed to the SERVQUAL scale to reflect the differences found in the banking services sector (Avkiran, 1994 and 1999, and Carman, 1990). Another finding was that the ratings of the scale items needed adjustments (Lewis, *et al.,* 1994).

6.3.1.2 Interviews

A qualitative approach is needed where the research questions are largely exploratory and at a pre-hypothesis stage. In such instances, interviews are used as a method to measure attributes that would affect purchase intentions in banking services. Some of these studies are listed in Table (6.2, next page).

Title of Study	Author(s)
Service Quality in the Banking Industry: An Assessment in a Developing Economy	Angur, M.G., Nataraajan, R. and Jahera Jr., J. S. (1999)
Another Look Into the Agenda of Consumer Satisfaction: Focusing on Service Providers' Own and Perceived Viewpoints	Athanassopoulos, A. D. (1997)
Developing an Instrument to Measure Consumer Service Quality in Branch Banking	Avkiran, N. K. (1994)
Quality Consumer Service Demands Human Contact	Avkiran, N. K. (1999)
Modelling Consumer Choice of Distribution Channels: An Illustration From Financial Services	Black, N., Lockett, A., Ennew, C., Winkihofer, H. and McKechnie, S. (1999)
Modelling Consumer Choice of Distribution Channels: An Illustration From Financial Services	Black, 'N., Lockett, A., Ennew, C., Winkihofer, H. and McKechnie, S. (1999)
Measuring Service Quality - A Re-examination and Extension	Cronin Jr., J.J. and Taylor, S.A. (1992)
Service Quality in Retail Banking: The Experience of Two British Clearing Banks	Newman, K. and Cowling, A. (1996)

Table (6.2): Studies utilising interviews

The above collection contains three different variations on interview questioning, i.e. unstructured, semi-structured and structured interviews. While the majority of the studies relied on individuals, Black, *et al.*, (1999) used focus group interviews. A very large-scale structured interview process was undertaken by Athanassopoulos (1997), with 270 respondents. Angur, *et al.*, (1999) describes his methodology as "interviews" but in effect they were really a verbally administered close-ended questionnaire.

6.3.1.3 *Case Studies*

Studies based on a case study approach to measure attributes such as service quality and consumer satisfaction, and how these would affect purchase intentions in banking services and are listed in Table (6.3).

Title of Study	Authors
Case Study: Service Quality, Business Process Re-engineering and Human Resources: A Case Point?	Newman, K., Cowling, A. and Leigh, S. (1998)
Barclays Life Consumer Satisfaction and Loyalty Tracking Survey: A Demonstration of Consumer Loyalty Research in Practise	Drake, C., Gwynne, A. and Waite, N. (1998)
Service Quality in Retail Banking: The Experience of Two British Clearing Banks	Newman, K. and Cowling, A. (1996)

Table (6.3): Case studies

Newman *et al.*, (1998) takes a triangulatory approach, based on company documentation, interviews with company managers, academic and practitioner conference papers as well as independent stockbroker reports. Company documents include annual reports and accounts, some service quality documents, including brochures on service standards and some of the quality club publications, some human resource strategy documents related to employee opinion surveys, a template for the new roles and job specifications generated by the process redesign and service quality measures and some issues of the staff newspaper. Personal interviews were conducted of varying lengths with bank personnel at varying positions in the hierarchy. Since most banks tended to be geographically distributed, this study attempted to generate a local and regional perspective, thereby increasing representativeness and reducing biases in the interpretation of evidence generated by the centre.

Other studies that adopted the case study approach combined it with large-scale surveys (Drake, *et al.*, 1998), so that the information gathered from surveys can undergo tests of a quantitative nature. Another approach taken by some researchers (Newman and Cowling, 1996) is to combine the analysis of company documents and with semi-structured interviews.

6.3.2 Data Analysis

The previous section examined data collection methodologies employed in previous empirical research work. This section examines how data collected by the above methods were analysed.

Without exception, all studies where quantitative data is involved carry out rigorous statistical tests. Sample biases are tested by carrying out Pearson's x2 tests (Mols, 1998b). Items of measurement are investigated for reliability by using Cronbach's alpha (a) (Mols, 1998b, Bloemer, 1998). A figure larger than 0.70 is deemed to be sufficient to conclude that the data gathered is reliable (Mols, 1998b). If the construct being measured is multidimensional, as it is the case with service quality, consumer satisfaction, etc. it is important that the coefficient *a* is computed separately for each of the dimensions to ascertain the extent to which items making up each dimension shared a common core (Parasuraman, *et al.*, 1988). Further refinements are accomplished by the principal axis factoring procedure. Parasuraman *et al.*, (1988) rotated factor solutions both orthogonally and obliquely to allow for inter-correlations among the dimensions and to facilitate easy interpretation.

Mols (1998b) carries out Kendall's tau-b (t) zero order correlation coefficient analysis, because the method has the following advantages; (1) it is a non-parametric measure of association between variables, hence it only requires an ordinal scale, (2) it eliminates the effects of outliers, (3) does not require a knowledge of the distribution of the variables in order to be able to interpret the probability tests. He also reports that Pearson's product moment correlation and Spearman's rho correction produce approximately the same results as the use of Kendall's tau-b. McDougall and Levesque (1994) use a cluster based, or post-hoc segmentation analysis approach to identify similar segments of respondents. An additional measure to highlight segments is the discriminant function, i.e. the calculation of the standardised discriminant function coefficient.

The number of research studies that use regression analysis are in the minority. Lewis *et al.*, (1994) use simple regression analysis, while Bloemer, *et al.*, (1998) utilise multivariate regression analysis for hypothesis testing.

Where the research questions attempt to explore qualitative details, no significant statistical analysis methods are used. Where the interview questions are structured, and the responses can be quantified, data analysis is carried out using methods similar to those described under questionnaire data analysis above. For example, Angur, *et al.*, (1999) carry out confirmatory data analysis using LISREL 8, to test for generalizability of the five-factor conceptualisation of service quality. They also conduct exploratory factor analysis using the OBLIMIN oblique factor analysis to test for service quality multidimensionality.

6.3.3 Methodological Justification

It was seen that empirical research in traditional banking services utilised three primary data collection methodologies, questionnaires, interviews and case studies. The above section examined how the data collected by these methods were analysed. This section examines how the researchers justified the selection of a particular data collection methodology.

High reliabilities and internal consistencies are necessary conditions for construct validity. Construct validity needs to be built upon solid conceptual and empirical criteria to be considered good. Parasuraman, *et al.*, (1994) postulates that construct validity is supported by face validity, a subjective criterion reflecting the extent to which scale items are meaningful and appear to represent the construct being measured. The argument being that scale measures were developed after a literature survey, pilot study and consultative panels. Similarly, many studies investigated face validity by investigating the comprehensibility and the ease of understanding of the questionnaire items during the questionnaire design phase by evaluating the consistency of the questions' content to different people.

Mols (1998b) ensures discriminant validity by investigating each single item in the questionnaire to see if its item-total correlation *is* lower *than* its correlations *with any* other variable in the study. On the other hand Carman (1990) and Brown, *et al.*, (1993) state that discriminant validity is indicated if the factors, and their items, are truly different from one another. Or in other words, discriminant validity is the extent to which a variable "discriminates" itself from other measures. A common problem in this case is how the reliability of measures affects discriminant validity. Low measure of reliability attenuates correlations between constructs. Thus a measure with low reliability may appear to possess discriminant validity simply because it is unreliable. Discriminant validity can be assessed by looking at the correlations of alternative measures of service quality with other research variables such as "satisfaction" and "future usage behaviour". In other words, service quality scales have high discriminant validity if they correlate more highly with each other that then do with other research variables (Churchill, 1976; Cronin and Taylor, 1992; Parasuraman, *et al.*, 1994). Concurrent validity refers to the extent to which instrument scores are associated as hypothesised with other conceptually-related measures.

Assessing a scale's content validity is necessarily qualitative rather than quantitative. It involves examining two aspects; (1) the thoroughness with which the construct to be scaled and its domain are defined and (2) the extent to which the scale items represent the construct's domain. The scale's validity can also be assessed empirically by examining its convergent validity. Convergent validity pertains to the extent to which scale items assumed to represent a construct do in fact "converge" on the same construct (Parasuraman, *et al.*, 1994). It is tested through subsets of questions and by calculating correlations with the alternative measures used.

Dimensionality of a scale (the actual instrument) can be measured by a confirmatory factor analysis (Cronin and Taylor, 1992). If the factor analysis is not confirmed, the unidimensionality of the items can be assessed. For this purpose, a factor analysis is carried out, using OBLIMIN factor rotation, and a coefficient alpha is more than 0.70 confirms that the scales are indeed unidimensional. Athanassopoulos (1997) contends that if alpha values exceed 0.70, scales appear reliable and demonstrate high internal consistency.

Consistent factor structures of a set of measures across several independent samples provide support for trait validity (Parasuraman, *et al.*, 1988).

Stepwise regression analysis is used to assess the predictive (or concurrent) ability of the alternative measures of service quality.

What is clear from the above analysis is that it is very important to select an appropriate data collection methodology to address a particular research question. Thereafter it is equally important to ensure that the data so collected is analysed in a robust manner. Last but not least, it is important to justify the selection of a particular data collection methodology. The task at hand is to explore the possibility of adopting one of these data collection methodologies to this research.

6.4 Adaptation of thcsc methods to an Internet Environment

There are many potential advantages offered by the Internet for data collection research methodologies. Some of these include greater speed, lower cost, improved accuracy in encoding data and the ease of obtaining international samples (Johnston, 1999). However, Pincott and Branthwaite (2000) assert that these claims are seldom backed up by detailed, comparative analysis of all the costs (preparation of data collection tools, collecting data, analysis, etc.) and caution should be exercised in assuming that the benefits are universal and apply to all forms of research. Others such as Kent and Lee (1999) and Schillewaert, *et al.*, (1998) subscribe to this view by contending that, in emphasizing the opportunities and potential of the Internet, reviews tend to gloss over the problems and limitations, which include poor response rates and unrepresentative samples (that lead to inaccuracy in assessing consumer attitudes and reactions).

6.4.1 Attitudinal Research in an Internet Environment

In particular, as far as attitudinal research is concerned, as it is the case in research carried out earlier in traditional banking services and examined above (in section 6.3.1), the relationship between the respondent and the researcher is very important. The nature of the relationship between the researcher and respondent through the Internet may be different, for the relationship is no longer a personal one in the way that is achieved compared with face-to-face or telephone contact. These effects can be observed in a number of studies.

1. Internet respondents can be more self-centred and self-absorbed, though there can be positive effects from this. According to Mehtra and Sivada (1995), electronic responses to open-ended questions are up to three times longer, contain clarifications and explanations, and have more candour and are less inhibited. However this may depend on the topic being of interest and importance to the respondents.
2. Differing findings are presented by Scholl, *et al.*, (1999). They contend that respondents are likely to submit shorter and more superficial responses, in comparison with face-to-face qualitative interviews.
3. Adriaenssens and Cadman (1999) noticed that following an unplanned telephone contact between the moderator and respondents in moderated email groups (MEGs), the tone and nature of the interactions changed. Respondents became more friendly and made jokes in their emails, which in turn could affect the quality of responses.
4. There is a tendency for individuals to adopt extreme positions in the Internet. Internet relationships can be polarized, either being very serious, formal and task focused or playful, surreal and fantastic where respondents revert to their cyber personalities (Pincott and Branthwaite, 2000).
5. Pincott and Branthwaite (2000) assert that it can be difficult to interpret responses in the vacuum created by anonymity without knowledge something about the age, gender, lifestyle of the respondent to give the response context and meaning. This apparent lack of identity can also result in an individual submitting multiple responses.

These highlight some of the potential implications that may be of concern. What is clear is that there are equal threats to the measurement, validity, reliability, generalizability and relevance of findings of attitudinal research carried over the Internet and highlight the need for alternative methods.

6.4.2 The Need for an Alternative Methodology

Section (6.2) illustrated how Internet banking services differ from traditional banking services. This section also highlighted that in Internet banking services, the need to carry out attitudinal research is somewhat reduced, because many attributes of the service can be evaluated using objective criteria. However, it was seen that all previous empirical research in traditional banking services relied primarily upon attitudinal research methodologies for data collection. Methodological rigour was examined by analysing how well the researchers defended their chosen methodologies. Section (6.4.1) argued that carrying out attitudinal research on the Internet might be problematic.

Therefore, it appears that there is an argument for an alternative methodology that avoids or minimises the disadvantages of attitudinal research. Hence there is a need to examine a number of issues (some of which have been already highlighted) that may strengthen this argument and aid in the determination of a suitable alternative research methodology.

1. As discussed above, there is widespread disagreement on how the respondents might respond, in moving from an interviewer-administered, face-to-face questionnaire to self-completion questionnaire with no research personnel to help and advise (Strauss and Frost, 1999). Consequently results tend to be treated with caution.
2. There is a whole range of issues that pose enormous problems for a given methodology. When respondents are asked about some aspect of service quality, the respondent's PC including hardware specification, configuration and installed software, quality of the connection to the Internet, etc. will have a major influence on responses. It is very difficult, if not economically impossible, to maintain the required uniformity, for generalisable results.
3. Internet banking has been in existence only for a short time. The earliest Internet banking service, BankNet Electronic Banking Service, claims to have started operations in January 1995. However, this service attracted few users, and it is never quoted by the popular press in their periodic surveys of Internet banking services. Additionally, an examination of its Web site reveals that the service does not seem to have progressed beyond the level of technology that was available in 1995. For instance, the author first came across the Web site in early 1998, and a beta version of the online cheque writing facility that was present then had not been upgraded to a fully functional version even by early 2000. Of the more popular services, Royal Bank of Scotland was the first to launch a transac-

tional Internet banking service in June 1997. Six of the fifteen banking services included in this research commenced operations in the latter half of 1999. All this means that time-series observations are so far unavailable, for they would not yield sufficient number of observations. Meaningful time series studies cannot be undertaken until the industry matures.

4. Data gathered directly from Web site observations must appreciate the ephemeral nature of Web sites and the fact that they are modified frequently, without leaving a trace of preceding offerings. This means that potential consumers are unlikely to keep track of changes and are unlikely to remember the evolving nature of the sites. Hence, attitudinal surveys using consumers as respondents at two different times will not result in replicable outcomes. This strengthens the case against carrying out attitudinal surveys on consumer perception on Internet banking services delivered over the Web. Any method adopted must take into account the dynamic nature of Web sites, therefore every effort was made to store observations to aid subsequent replicability.
5. It was argued that Internet banking services contain significantly higher levels of search qualities in comparison to traditional financial services. Additionally, attributes such as experience qualities are relatively easily acquired in Internet banking services. It was further argued that in certain aspects Internet banking services are more akin to a physical product than a traditional service product. Therefore it is possible to compare different Internet banking services by evaluating them side-by-side, facilitating more objective measurement criteria and consequently reducing the emphasis on attitudinal research methodologies. This makes the data collection process much easier and more objective.
6. Section (3.6.1) examined dimensions of an Internet banking service encounter. It was found that many of these dimensions can be evaluated by objective criteria, reducing the importance of individual character traits. This is an additional reason to place less emphasis on attitudinal research methodologies.
7. Due to the nature of Internet businesses, especially the value attributed to the captive consumer base, it is unlikely that even one Internet bank will give researchers email access to its consumers for the purpose of carrying out attitudinal research. In the UK, statutory data protection laws make this process even more difficult. Already, many Internet banks solicit the views of their customers, through various Web-based and traditional instruments, for the

continuous development, presentation and delivery of their Internet banking service products. As such, the findings of these attitudinal surveys are highly commercially sensitive. Additionally, the primary target group of consumers that this research is concerned with is individuals that are yet to purchase Internet banking service products. Therefore, customers that are already consuming Internet banking products from a particular bank or a number of banks, for which the banks would be a source of contact, would not constitute the population from which a representative sample could be drawn.

8. Related to above, the number of customers will be important for the Internet banks on several grounds. Firstly, in the 'dot.com' economy the number of 'eyeballs', or the number of customers, is one of the primary determinants of the level of success as perceived by capital market investors. Secondly, as explained earlier, significant 'bandwagon' effects (Moschandreas, 1994) provide a motive for banks to over-estimate the true size of their base of active consumers, for example, through the continued inclusion of inactive consumers. This makes triangulation (Yin, 1993) particularly desirable in the context of auditing banks' own sales estimates. In the absence of third party audits, the cooperation of Internet banking institutions would not aid in the unbiased estimation of user volumes.
9. An important aspect of the Internet for users is privacy (Strauss and Frost, 1999). According to Hoffman *et al.*, (1997), 94 percent of Web users have refused to provide information at a Web site. Perhaps one of the most significant findings of their research is that 40 percent of users give false information. Such findings question the validity of online research conclusions.
10. As stated in the introductory chapter, in a typical household the person who buys banking products may not be the only person who consumes it and similarly the decision of exactly what to buy may be taken in consultation with other members of the household. As a result a simple questionnaire type of instrument may not capture the influences of all concerned on a purchase.

Many Internet services, as shown in the case of Internet banking services, involve the routinization of many aspects of service provision. This reduces the difficulties involved in the measurement of many aspects, and as argued above reduces the dependence on attitudinal research methodologies as it is the case with traditional banking services.

On the other hand, Internet services offer enormous new opportunities for carrying out research. Although direct surveys of consumers are virtually ruled out, it is much easier for researchers to simulate the consumer's search for and perception of an Internet banking service by simply logging on to a Web site and making observations. At this early stage in the development of Internet banking it is not possible to test relationships between web site properties and purchases over time, and so it is only possible to perform the first step in this commercial and methodological environment. The next sections describe the proposed new methodology.

6.5 Mystery Shopping

A number of articles on 'mystery-shopping' have appeared in the trade press (Miles, 1993; Newton, 1994; Cramp, 1994; Burnside, 1994; McNeil, 1994; Dwek, 1996). However published academic research has been limited to a small number of papers (Wilson, 1998). Dawson and Hillier (1995) discussed the ethics of mystery-shopping, Morrison, *et al.,* (1997) researched the cognitive processes of shoppers, Wilson (1998) examined the role of mystery-shopping in the measurement of the service delivery process across a number of service sectors, and the remainder have tended to focus on its use in specific sectors, for example, Leeds (1995) and Dorman (1994) examined the banking sector.

Therefore, there is a need to examine this methodology from a broader perspective in order to evaluate it more carefully. In this context, an illustration of the background to the development of the mystery-shopping methodology by expounding its foundations in participant observation will be of use.

6.5.1 Participant Observation and Mystery Shopping - The Foundations

The use of participant observation, also called situation research and mystery-shopping, where the researcher interacts with the subject or subjects being observed stems from the field of cultural anthropology (Wilson and Gutmann, 1998). Anthropologists would take part in a tribe's daily life in order to understand the norms, attitudes and behaviours that were neither documented nor communicable via language. Anthropology tended to use unstructured observation with researchers immersing themselves in a situation and discovering 'reality'. Covert participant observation, where the observation is undertaken in a concealed manner to ensure that the experience is natural and not contrived for the sake of the observer, has also been used in the social sciences, for example in the study of race and sex discrimination in recruitment procedures (Esmail & Everington, 1993).

Additionally, the Consumers' Association in the UK have been using covert participant observation in their tests of service performance since 1960 (Brown, *et al.*, 1997). Even Government Departments have used the technique to monitor areas such as postal deliveries or the testing undertaken in MOT test centres (garages which test the roadworthiness of cars, COI, (1996)).

Participant observation has tended to be used by each of these bodies as it allows the researcher to overcome some of the potential weaknesses of interviewing and survey research. These are highlighted by Friedrichs & Ludtke (1975).

1. There is often a discrepancy between real and reported behaviour. Occasionally statements are made in interviews which are not in accordance with the factual behaviour of the interviewed persons.
2. Often facts are brought to light by means of natural settings only. The interviewee is not conscious of them and they are therefore not easy to get at by questioning.
3. The verbal capabilities of the interviewed person can limit the quality and quantity of information gathered.

In a services context, observational methods are able to provide information of the service experience as it unfolds (Grove and Fisk, 1992). In particular, participant observation helps to develop a richer knowledge of the experimental *nature of services*. The participant can identify dimensions of the service encounter unlikely to be discerned by a distant or non-participant observer. Concealment of this observation, although raising ethical issues can ensure that the experience is natural and not contrived for the sake of the observer.

In terms of ethics, observing people without their knowledge may violate their rights to privacy and freedom from exploitation (Jorgensen, 1989). However, services are often performed in public settings where their delivery can often be observed by members of the public other than the specific obligation to the service provider in a concealed observation than in normal everyday life circumstances.

Mystery shopping is a clear example of concealed participant observation in a public setting. However, it differs from original anthropological approach to observation in terms of its structured and systematic format. Cultural anthropology tended to use unstructured observation with researchers immersing themselves in a situation and discovering 'reality'. Whereas mystery-shopping uses a structured approach of checklists and codes to gather and measure specific information about service performance in everyday conditions (Grove and Fisk, 1992).

The need for specific performance information stems from the increasing emphasis on service performance by service mangers. Whilst head office staff and senior management invariably set service standards, the task of delivering these standards falls to individual staff member who face customers. Variations in service performance can have a major impact on customer satisfaction. This is emphasised by Bateson (1995) who suggested that the customer's experience during the delivery of a service is as important to customer satisfaction as is the benefit that the service provides. Gronroos, (1983) in his paper on 'interactive marketing' also stressed that 'how' you perform a service is often as important as 'what' you perform.

Gronroos (1982) as well as Langeard, *et al.*, (1981) identified the interactions that occur during the performance of a service.

1. 'The consumer versus the physical/technical resources' - the interaction between the consumer and the exterior and interior of a bank or the computer systems of a bank.
2. 'The consumer versus the contact personnel' - the quality, service mindedness and customer sensitivity of the contact personnel.
3. 'The consumer versus consumers' - the active involvement of the consumer in the production process and the influence consumers have on each other.

The importance and nature of each of these interactions varies depending on the service being provided and the consumers being targeted. For example, Gronroos (1982) asserted that for some services, no human resources are involved in service provision, only physical resources are needed. However, the successful management of service delivery involves the co-ordination of these various interactions and supporting them with the relevant procedures, personnel and systems (Segal-Hom, 1988; Berry, 1983).

A critical part of the procedures and systems is the setting of service standards (Berry, *et al.*, 1985, 1988; Johnston, 1988; Lovelock, 1988; Gronroos, 1981). As mystery-shopping has the potential to measure service performance against pre-set quality standards it should play an integral part in this evaluation by providing the measurement of performance (Leeds, 1995; Armstrong and Arvanitis, 1996; Brown, *et al.*, 1997).

6.5.2 Mystery Shopping and its General Uses

It is apparent from the above that, rather than asking customers or passengers about the service they receive, mystery-shopping, the name given to participant observation in the commercial sector, enables researchers to experience and view the tangible and intangible elements of a service for themselves. Additionally, mystery-shopping uses a structured approach of checklists and codes to gather and measure specific information about service performance in everyday conditions (Grove & Fisk 1992). The information obtained is more detailed than that gathered through customer satisfaction surveys and provides evidence on whether activities do or do not happen rather than gathering broader opinions about the service experience (Wilson 1998). It is aimed at measuring an organisation's service performance against a number of pre-set quality standards (Leeds 1995; Armstrong & Arvanitis 1996). Emerging academic and other literature shows that mystery-shopping measurements tend to be used for one or more of the following purposes.

1. To act as a diagnostic tool identifying failings and weak points in an organisation's service delivery (Wilson, 1998; Wilson and Gutmann, 1998). As a diagnostic tool, respondents use mystery-shopping to track key elements of their organisation's service delivery process and help identify where capital, technical and human resource requirements to be allocated. Decisions on actions relating to internal processes and competencies are therefore supported by information from a customer viewpoint rather than being based solely on an operation or management perspective. This customer viewpoint differs from the measures provided by customer satisfaction surveys, as managers can look behind the aggregate measures and identify specific service failings, on particular locations, on certain times, responding to identifiable types of customer or enquiry.
2. To encourage, develop and motivate service teams by linking with appraisal, training and reward mechanisms (Dorman, 1994). Definitive information from mystery-shopping is also seen as being essential for appraising, developing, motivating and rewarding service teams and potentially individual service personnel. In terms of motivation, mystery-shopping results can be used to reward those service teams who are performing well against the set service standards. Financial rewards and incentives are seen as becoming more common in almost all sectors but particularly in the financial services sector (Wilson, 1998; Dorman, 1994).
3. To assess the competitiveness of an organisation's service prov1s10n by benchmarking it against the offerings of others in an industry (Wilson, 1998; Wilson and Gutmann, 1998). Mystery shopping can also be used as a bench mark comparison with the service performance of competitors. Where the research is undertaken on a syndicated basis, this is relatively straightforward but where an

organisation commissions a mystery-shopping programme which includes an assessment of the competitors, certain ethical issues become important (Dawson and Hillier, 1995). The Market Research Society (MRS) Code of Conduct specifies that mystery-shopping should not involve an unreasonable amount of time or expense on behalf of the organisation being researched.

At this point it is worthwhile to examine the use of mystery-shopping in the banking services sector.

6.5.3 The Development and Use of Mystery Shopping in the Banking Sector

Mystery shopping in banking has evolved over the years, and its origins are in the US (Dwek, 1996). The US continues to be the biggest user of, hence it is useful to briefly illustrate the development and use of mystery-shopping in the US.

6.5.3.1 Mystery Shopping in the US

Mystery shopping became established during the 1970s, with approximately 25 percent to 35 percent of all banks with over $300 million in deposits conducting some type of mystery-shopping program (Leeds, 1995). What prompted the growing interest in mystery-shopping in the '70s was the realization by marketers and bankers of the importance of developing a sales culture (Reed and Miles, 1995). Because sales professionalism became increasingly important, a device had to be developed to monitor sales skills, as well as changes in service behaviours in the sales culture. Mystery shopping began to be used as a monitoring device for sales culture development, specifically for tracking sales behaviours and skills (Leeds, 1995; Reed and Miles, 1995).

This phenomenon then led to the use of mystery-shopping to not only monitor, but to motivate performance, set goals or standards and reward performance. Some of the more progressive and sales-oriented banks began rewarding employees based upon the sales achieved (Dorman, 1994). In the 1980s, the financial industry's new catch phrase was "service quality" and, once again, mystery-shopping (along with consumer and customer satisfaction surveys) was used extensively for evaluating, monitoring and motivating performance. It was the combination of these two research methodologies that changed the basic mystery-shopping methodology to one of a predictor of customer satisfaction. By determining what customers' wants and needs are and what satisfies customers most, checking for and reinforcing specific behaviours can be built into the mystery shopper program (Leeds, 1995; Dwek, 1996)

While the proponents of the methodology were deriving benefit from the mystery-shopping technique, it was not without critics. Dwek (1996) states that mystery-shopping used to be regarded as corporate spying. It was seen as an underhand activity, probably prompted by management accountants advising clients to axe hundreds or thousands of jobs. It was, in short, an excuse for firing people (Dwek, 1996)

6.5.3.2 Mystery Shopping in the UK

In the UK, mystery-shopping is the smallest area of specialisation covered by major research companies (Dawson and Hillier, 1995). However, in a survey conducted by the MRS in 1995, it was found that the use of mystery-shopping as a research methodology is gaining widespread acceptance (Dawson and Hillier, 1995). There are few published figures on the scale of mystery-shopping activity, Wilson (1998) gave an estimate of between £20-30million per annum and Dawson and Hillier (1995) identified 187 agencies providing this form of research.

Apart from using mystery-shopping as a part of performance monitoring programmes, competitor mystery-shopping is relatively widespread (Dawson and Hillier, 1995). However there are also a number of areas that lead to tension among companies with regard to competitor mystery-shopping. In a survey carried out by the MRS in 1995, the following were cited by the respondents to be problematic issues in competitor mystery-shopping: mystery shoppers take up too much time, should not do anything out of ordinary, should make a purchase, it is ethically wrong to place orders and then cancel, the taping of conversations is not acceptable unless prior permission is gained, etc. (Dawson and Hillier, 1995). As a response to these problems, and also to reduce the costs of research for individual organisations, syndicated studies involving a number of clients are also in operation. The syndicated studies enable the client organisation to assess their performance relative to industry norms and comprise groups of organisations. The financial services in particular have been engaged in a syndicated shopping since 1992. Subscribers such as the Midland, NatWest, Royal Bank of Scotland, Bank of Scotland, Bradford and Bingley, Nationwide, Yorkshire Building Society and Leeds Permanent get comparisons on their own and competitors' performances on a whole range of services (Miles, 1993).

Increasingly there is a tendency to deliver services through a multitude of channels. This development began with the increasing use, connectivity and falling costs of the now ubiquitous telephone. During the last decade, the delivery of financial services in particular, including banking service products, insurance services, brokering services, etc. have become widespread. Mystery shopping has been adopted to this medium, and according to NOP research, there is a move towards conducting more mystery-shopping surveys over the phone (Dwek, 1996).

The task at hand is to propose an innovative adaptation of the mystery-shopping survey methodology to Internet banking services. In order to achieve this objective the next section of this chapter is organised as follows. Firstly, the decision to adapt mystery-shopping methodology to Internet banking services is justified. Secondly, an overview of how the mystery-shopping methodology is adapted to Internet banking services is presented. The next chapter builds up on this section by operationalising the mystery-shopping research methodology. Thirdly, strengths and weaknesses of adapting the mystery-shopping methodology in Internet banking services are examined. In particular these aspects are compared and contrasted with attitudinal research methods (such as, questionnaires administered to consumers and interviews of consumers) discussed in the preceding sections of this chapter. And finally, how the mystery-shopping research methodology measures up to criteria governing good research design are considered.

6.6 Mystery-Shopping in Internet Banking Services

The discussion so far has shown that in the traditional service provision sectors, mystery-shopping as a research methodology has been adopted to achieve one or more of the following objectives. Firstly, it can be used as a diagnostic tool in identifying failings and weak points in an organisation's service delivery. Secondly, it can assess the competitiveness of an organisation's service provision by benchmarking it against the offerings of others in an industry. Thirdly, it can be used to encourage, develop and motivate service teams by linking with appraisal, training and reward mechanisms.

In an Internet banking service provision environment, the primary consumer channel is not an employee of the bank and a consumer. Rather it is a consumer interacting with the Internet interface, the Web pages of the bank's Web site. The consumer establishes contact with the bank via a communications network, each Web page that the consumer visit means that they are downloading all or most of the information on that page into their computer. As stated earlier, since Internet banking services are still in their infancy, in addition to this primary interface, all banks make use of the telephone as a secondary support channel. Additionally, those banks that do have branch network, also rely on the branch network as a third channel for the service offering. Therefore, although primary delivery channel is largely automated and dependent on the machines, the second and the third channels are dependant on the humans manning the respective ends of the channels.

A strong argument in favour of mystery-shopping in Internet banking services is that subsequent experience over the Internet is likely to be identical to trial experienced through browsing the web site and the demonstration component. A trial is not a common option in the case of traditional banking services. It may be possible to try one service episode in certain traditional services to establish whether one will purchase a full service or a sequence of episodes, but there is no guarantee that the trial experience will be identical to the actual service experience.

Gabbot and Hogg (1998) observed that in a physical product a trial is exactly equivalent to the actual product. In services there is no guarantee that the rest of the service bears any relation to the experience of a mystery-shopping trial. This is because a terrestrial service is intangible and the participation of any other individual gives no guarantee of a repeated performance hence mystery-shopping observation may be an unreliable as a source of information. It was noted that as a consequence, a number of authors suggest that consumers look towards the experiences of others (Murray, 1991). Further, Zeithaml (1981) postulates that experience information of the service prompts reliance upon word of mouth sources as they are perceived to be more credible and less biased. In Internet banking services, where the majority of operations are automated, there is a very strong likelihood that a consumer may receive rest of the service exactly like the trial. A trial involves experiencing the Web site and all its components, including the demonstration component and communication methods. In these novels circumstances the consumer experience the trial and acquire information, as a consequence may regard it to be the most credible and least biased source of information. Hence this is one of the strongest arguments for using mystery-shopping in determining consumer purchase intentions in Internet banking services.

An argument can be put forward that in assessing the purchase intentions of consumers, all three channels (the Internet, the telephone support channel and branches) need to be evaluated. However, this may create other difficulties that may influence the findings. Firstly, it is very difficult to carry out a compressive mystery-shopping data collection exercise on all three channels, for it will consume a vast amount of resources, both in terms of time and money. Secondly, there are ethical considerations as noted by Dawson and Hillier (1995). And finally, and most importantly, the objective of this research is to evaluate factors influencing purchase intentions in Internet banking services. By evaluating other channels there is a danger that the research findings may be diluted. For instance, a particular Internet banking service may have an outstanding telephone support, but its primary channel, the Internet, is poor in comparison to competitors. The same could be said of an Internet banking service with an excellent branch network providing excellent service, but with a poor primary channel. Hence the mystery-shopping data collection exercise is limited to the evaluation of an Internet banking service delivered over the primary channel.

6.6.1 Adapting Mystery-Shopping Methodology to the Internet

The objective of this section is to present an overview of how mystery-shopping methodology is adapted to collect data on characteristics of Internet banking services that influence consumer purchase intentions. This exercise may be construed as the operationalisation of the research methodology. Admittedly, this illustration of the adaptation process lays the groundwork for the operationalisation of the methodology. The operationalisation is discussed in more detail in the next chapter.

A thorough literature review, both in academic journals and in trade journals, failed to produce any previous research relating to mystery-shopping on the Internet. In Internet banking services there is no direct human intervention in the service delivery process, therefore services delivered from one consumer to another will be identical. The only intervening factor that could alter the homogeneous nature of the service is the performance of the Internet itself. As stated earlier, the performance of the Internet will depend on the quality and capacity of communication devices, the computer and peripherals used to gain access to the Internet and the amount of Web traffic at the time in question. This growing commodification of banking services delivered over the Internet means that it renders the researchers task much easier, since the evaluation process can be largely objective (Lowe and Kuusisto, 1999). This is because purchase decisions on commodities are largely determined by their technical qualities. An Internet mystery-shopping methodology offers the opportunity to gather information quickly and relatively cheaply. The service is offered remotely, so there are no transportation costs or delays. A researcher is able to carry out a desk-based research. Further, It offers total anonymity, and is a true "mystery" shopping scenario with the service provider not being aware that they are being surveyed.

The data gathered by this method are largely superior to data gathered through other research methodologies, since many measurements can be objectively measured and do not have to rely on the competence of the researcher. The quality of the data gathered is more dependant on the quality and comprehensives of the instrument used to record and collect information. Additionally, many freely available tools on the Internet, such as shareware software applications, can be used to enhance the measurement criteria.

Most Internet banking services rely on two components of the Internet, i.e. the WWW and email. The delivery of Internet banking service products takes place over the WWW. The Web offering may operate on a HTML application or on a Java application. Email is used primarily for bespoke and confidential communication between the consumer and the bank. This email could be encrypted to enhance the security level. A secondary use of email is to distribute information cheaply to consumers, these include, periodic newsletters, new product offering, changes of terms and conditions of products, etc. Additionally, when a prospective consumer logs on to the banking Web site, in addition to the general information about the banking service that is available on the Web site, there is an increasing trend to provide a demonstration of how the service would perform once such a prospective consumer becomes a customer of the bank. This demonstration service may run either as an HTML application or as a Java application. Hence, an Internet banking service means the totality of all these interrelated components and applications working in tandem.

In designing a mystery-shopping research methodology it is imperative that all the elements of the Internet banking service experience described here be included in the evaluation. Internet delivery channel is further divided into three sections, i.e. the WWW (or the Web site), the demonstration component and the email component.

Data collection instruments for each of these sections will consist of a series of scorecards with pre-determined evaluative criteria. The evaluative criteria within the scorecards are in tum are related to the hypothesis generated in the previous chapter. These will be explained in more detail in the following chapter.

6.6.1.1 *The Web Site*

As explained above the WWW constitutes the primary channel through which Internet banking services are delivered. It was also stated that there could be two types of applications (HTML and Java) running on the WWW. The Web pages can operate either in 'pure' form, for example, HTML only, or Java only, or increasingly in a combination of HTML and Java. These applications deliver, among others, text in various positions and contexts, graphics (both still and moving images), colours, sounds, etc. on to the consumer's computer terminal. The proposed mystery-shopping scorecard enables the researcher to examine how all these attributes of the Web contribute to consumer purchase intentions.

6.6.1.2 *Demonstration Component*

The availability of a demonstration greatly enhances the consumer's comprehension of the Internet banking service experience. Increasingly the banks are providing such a facility within the Web site. As in the case with the Web pages, the demonstration or 'demo' as these are popularly referred to, run either as an HTML application or a Java application or in a combination of both. Some demos require the consumer to download a specific file (in most cases an 'executable' file) into their hard disk and then run (execute) it from the hard disk. Other demos may work as an integral part of the bank Website. Further, some demos require the consumer's computer to be equipped with additional plug-in software, apart from the standard browser software. Many of the Internet banking services Web sites are the result of substantial investment in terms of technological attributes and these are in place to enhance and deliver a potential consumer a 'true feel' for the actual operation of an account, before they become a customer. Hence it is important that the mystery-shopping research proposed has the ability to evaluate how all these attributes of the demo contribute to consumer purchase intentions.

6.6.1.3 Email/Query Application Component

There is a balance between the amount of information that can be provided on the WWW by the banks, both through their Web site and any demos that may be part of it. Providing too much information may lead to an extremely large Web site which may put off consumers, by making it difficult to find information. The maintenance costs of a large Web site can be comparatively expensive. In light of the 'limited' amount of generic information available on the Web site, it is very likely that consumers may have specific queries that need bespoke answers. An appropriate channel is required for this purpose. Many Web sites provide an email address, or a query application on the Web site, which a potential consumer may use. Replies to these types of queries are usually sent to the consumer's personal email account. Where such a service is provided, it must be evaluated, and it is proposed to extend the mystery-shopping instrument to cover this aspect as well. It is appreciated that all banks do not provide either of the types of query service as described.

6.6.1.4 Overall Evaluation

As reiterated elsewhere, the assessment of purchase intentions of Internet banking services by consumers encompasses more than the mere evaluation of the Web site, its components and the different channels. As discussed throughout chapters two to four, a potential consumer is also influenced by other factors.

6.6.2 Strengths and Weaknesses of TMS Approach

Mystery-shopping research methodology suggested above, has the following inherent advantages.

Strengths

1. In contrast to customer satisfaction surveys, the mystery-shopping approach is being used to measure the process rather than the outcomes of a service encounter (Wilson, 1998). The emphasis is on the service as it unfolds, looking at which activities and procedures do or do not happen rather than gathering opinion about the service experience (Miles, 1993). Hence mystery-shopping is a very suitable methodology (for innovative application to an Internet banking context) as it provides more objective measures for ascertaining purchase determinants.
2. The desk-based mystery-shopping of the Internet banking service Web sites and demos produce very detailed measures at a fraction of the cost of other methods, such as a customer survey. The significant variable costs are limited to the wage costs of a researcher and the Internet connection costs. Incidentally, at the time of writing the Internet connection charges in the UK are continually coming down with the prospects of un-metered Internet connection costs a very much a prospect in the near future.
3. The data gathered in this method is largely superior to data gathered through other traditional forms of mystery-shopping, since measurements rely less on the competence of the researcher.
4. Related to above, while some of the measures are subjective, freely available software tools on the Internet can be used to enhance measurements, and by extension enhance the objectivity of measures.
5. Unlike other methods of data collection, such as a customer attitudinal survey, the data collection process and the data entry process can take place simultaneously. Modem operating systems for computers can easily accommodate a number of applications running simultaneously, hence a researcher is able to collect data using a mystery-shopping instrument running on one software application and at the same time key in the data into a database which runs on yet another software application. Such an arrangement saves a considerable amount of time, money, and moreover it would mitigate tendencies for any data transfer errors. Further this process would speed up the entire research progress substantially. Therefore it would be equally suited for the academic and commercial environ-

ments, perhaps more so in a commercial environment where time pressures are greater.

6. Unlike in field research, desk-based research affords more opportunities for controlling the effects of the external environment. Additionally, such effects of the environment, if they do occur, they can be monitored and recorded. Where maintaining constant controls are impossible, for example login time, time spent on reading the information on Web pages, etc. can be monitored and fed into the database, so that allowances could be made at the analysis stage.
7. The respondent's PC including hardware specification, configuration and installed software, quality of the connection to the Internet, etc. will have a major influence on their responses, and hence the importance of maintaining uniformity. It is very difficult, if not practically impossible to maintain the required uniformity among a representative sample of respondents for the results to be generalised. This methodology avoids this aspect altogether for only one PC is used for the entire process.
8. Internet banking services is a dynamic sector, where processes, procedures, technology, products, etc. all change at a rate incomparable to the terrestrial banking services sector. Hence any research methodology employed to measure any aspect of Internet banking service must have an inherent flexibility at its core to be a credible methodology. The qualities of mystery-shopping qualify as a more suitable methodology on these grounds.
9. Related to the above, in commercial applications, where there is a need to conduct frequent assessment of competitors, Internet mystery-shopping can be used very effectively. The MRS code imposes limits on: frequent assessments of competitors in the terrestrial world. Such limitations include, ensuring a spread of mystery-shopping assessments over time; not all in one week/one day; not carrying out mystery-shopping exercises at busy periods, with mystery-shopping visits to individual outlets restricted to a maximum of once a quarter, etc. (Miles, 1993). In an Internet environment such limitations will not apply, since in practical terms, an Internet mystery-shopping survey will not consume competitor's resources.
10. It offers total anonymity, and is a true 'mystery' shopping scenario with the service provider not being aware that they are being shopped.

11. It also avoids the ethical considerations of a traditional terrestrial mystery-shopping methodology. The resources of an Internet banking service provider are wasted. Strictly speaking the mystery-shopping evaluation does add to the number of users using the servers of the bank, hence increasing traffic. However, the marginal increase is negligible when most of these back end servers are typically designed to handle 100,000 users at a given time.
12. Hoffman *et al.,* (1997) state that 94 percent of Web users have refused to provide information at a Web site, and moreover of those who do provide information, 40 percent of users give false information. Such findings question the validity of research conclusions arrived through Internet based surveys. The mystery-shopping methodology provides no opportunity for respondents to lie, for only a researcher is involved in the data collection.

Weaknesses

1. The entire outcome of the methodology depends on the quality of the data collection instrument. Therefore in designing the instrument, a great deal of work has to be put in reading the appropriate material to make sure that all that is needed to be measured is measured, for if a crucial factor is omitted it may yield spurious results.
2. Because the methodology does not solicit views from a sample of respondents, it lacks the richness or open-ended answers and suggestions.
3. Admittedly, some evaluative measures in the mystery-shopping data collection scorecards are subjective. While it is impossible to remove the subjective element, what can be done is limited to mitigate the degree of subjectivity of the findings by formulating appropriate measures that have objective underpinnings.

On balance, taking into account all factors, it is fair to state that mystery-shopping methodology represents an improvement over other attitudinal research methodologies for data collection. Caveats remain, but given the time and resources available for this research project it appears that to meet the criteria governing good research design.

6.3.3 How the Research Methodology Measures up to Criteria Governing a Goon Research Design

So far, a mystery-shopping research methodology has been suggested, its strengths and weaknesses discussed. This section examines how the proposed methodology measures up to criteria governing a good research design, with regard to measurement, structure, validity, reliability, rigour, and relevance of the measuring instrument.

6.6.3.1 *Measurement*

Nature and Levels of Measurement

Measurement is a procedure in which one assigns numerals, to empirical properties (variables) according to rules (Nachmias and Nachmias, 1993). The numerals which are the end product of measurement, are used for comparison, evaluation, and the assessment of relations between various properties. A numeral has no quantitative meaning, unless it is given one. Numerals that are given quantitative meaning become numbers; these enable the use of mathematical and statistical techniques for purposes of description, explanation, and prediction. A second concept that is used to define measurement is that of rules. A rule specifies the procedure according to which numerals or numbers are to be assigned to objects or events (Nachmias and Nachmias, 1993). Measurement, then, is the assignment of numerals or numbers to objects, events, or variables according to rules. Rules are the most significant component of the measurement procedure because they determine the quality of measurement.

There are four levels of measurement, nominal, ordinal, interval and ratio. The measurement levels used in this research will be limited to ordinal, for subjective assessments and ratio level for objective assessments of variables. The ordinal level of measurement levels used in this research will be limited to ordinal, for subjective assessments and ratio level for objective assessments of variables. The ordinal level of measurement is unique up to a monotonic transformation, that is, any order-preserving transformation does not change the information obtained (Nachmias and Nachmias, 1993). It does not matter what numbers one assigns to a pair of objects or to a category of objects so long as one is consistent (Curwin and Slater, 1997). It is a matter of convenience whether to use lower numbers for the 'higher ranks'. Almost all previous studies reviewed for this research have adhered to this principle of assigning lowest number to represent 'best', 'strongly agree', etc. Transformations that do not change the order of properties are permissible at the ordinal level. Accordingly, mathematical operations and statistics that do not alter the order of properties are also permissible (Hair, *et al.*, 1995; Curwin and Slater, 1997). For example, a statistic that describes the central tendency of ordinal numbers is the median. The median is not affected by changes in any numbers above or below it so long as the number of ranked observations above and below remains the same.

Most of the previous research on banking services, where subjective measurements such as consumer attitudes are measured using ordinal scales. There is an inherent weakness in this method. For example in the Parasuraman, *et al.*, (1985 and 1988) studies the *SERVQUAL* scale has no verbal labels for points 2 to 6 and so respondents may be over-using the extremes of the scales which do have verbal anchors because subjects have a tendency to focus on words (Lewis *et al.*, 1994). A further issue of concern is that the 7-point Likert scale used in *SERVQUAL* may camouflage subtle variations in consumer expectations and perceptions. For example, a consumer may feel his expectations lie between points 4 and 5, and record 5 (rather than 4.4) and their performance rating between 5 and 6 (rather than 5.6), therefore the recorded measurements show no difference between expectations and perceptions when the actual difference is 1.2. These slight variations become increasingly important if one begins to weight the items according to their importance.

These particular problems associated with previous studies are taken into account in the present research and a graphic positioning scale is used to assess the variables. A bi-polar semantic scale is incorporated rather like a Likert (poorest -best) scale. The scale, 1Ocm long is calibrated, at one-centimetre intervals, and provide a point of reference for the researcher from which they evaluate their subjective assessment of variables.

Best	1	2	3	4	5	6	7	8	9	10	Poorest

A slight drawback of the scale is that data entry takes longer, for the exact location of the evaluations must be carefully read off.

Measurement Error

Measurement procedures lead to the assignment of numerals, numbers of scores to properties. Differences in the scores can be attributed to two sources. One is the extent to which the properties exhibit real differences in the aspects of the property being measured. The other source of difference in the scores is the extent to which the measure itself or the setting in which it takes place influence the scores. In this case, the measures are showing unreal differences. Perfect measures reveal only differences between the properties. However, measures are seldom perfect and often reflect not only real differences but also artifact differences produced by the measuring procedure itself (Nachmias and Nachmias, 1993). Differences that are due to anything other than real differences are termed measurement errors. They are not real differences between the properties but differences due to the imperfection of the measuring procedure.

Nachmias and Nachmias (1993) identify five different sources of measurement error.

1. The scores obtained may be related to an associated attribute. For example in the evaluation of Web page design aspects may require an appraisal of the navigational aspects of the Web page to a certain level. In responding the this question the researcher will reflect not only on the design aspects, but also compare and contrast the navigational qualities. The effects of these associated attributes are measurement errors.
2. Measurement errors may result because of temporary differences in conditions. Every possible measure, such as carrying out the research at similar times on different days, etc. may be important. It is appreciated that this type of measurement errors cannot be totally eliminated, but steps can be taken to minimise them as much as possible.
3. Measurement errors may also result when different people interpret the measuring instrument in different ways. For this specific research work this is not a concern, as the author designed the instrument and he alone collected the data, hence there was no misinterpretation of the instrument. Nevertheless it is important recognise the need for the research instrument designer to explain the instrument fully to the researcher(s) collecting the data.
4. Differences in the setting in which the measure is used may contribute to measurement error. It is very unlikely that this type of measurement errors will creep into to the main component of the research. However, with regard to the telephone survey and the branch survey, the researcher's age, gender, sex, language, race, qualifications, know-how of technology, may influence the answers of respondents.
5. Differences in the administration of the measuring instrument may lead to measurement errors. Here again maximum care is taken to make sure that the survey takes place at comparable levels, and it is unlikely these errors will be of significance due to the nature of the research methodology.
6. Finally, distortions may occur in processing and analysing data. As identified earlier, data processing errors are also likely to be very low, for the author will collect the data and feed the data almost simultaneously. Care will be taken at the analysis stage to reduce errors.

Validity and reliability are concerned with techniques for reducing measurement errors.

6.6.3.2 *Validity*

The term validity is potentially confusing because it has been used in a variety of ways in the methodology literature (Weber, 1985). The problem of validity arises because measurement is indirect. Under such circumstances, there is some uncertainty as to whether the intended measure is really being measured. Validity is concerned with the question

> *"is one measuring what one intends to measure?"*
>
> (Nachmias and Nachmias, 1993, p.158).

To answer such a question, an attempt is made here to provide supporting evidence that the measurement methodology and the instrument do, indeed, measure what it appears to measure.

Several kinds of validity can be distinguished, each of which is concerned with a different aspect of the measurement situation.

Content Validity

There are two common varieties of content validity: face validity and sampling validity. Face validity is a subjective criterion reflecting the extent to which scale items are meaningful and appear to represent the construct being measured (Nachmias and Nachmias, 1993). The mystery-shopping scales satisfy face validity criteria in the following manner. Firstly, the scale measures are developed after an extensive literature review; with most of the scale items being cross-referenced (see appendix for explanatory notes on the Internet scorecard scales). Secondly, scale items are easy to understand, for the survey data collection instrument contain an additional column giving a detailed description of what they measure. Thirdly, a number of the scale items are objective measurements, thus overcoming the main problem of face validity as cited by Nachmias and Nachmias (1993), i.e. that there is no replicability procedures for evaluating the measuring instrument, because there is a reduction in the need for relying on subjective judgement of the researcher.

The primary concern of sampling validity is whether a given population of situations or behaviour is adequately sampled by the measuring instrument in question; that is, does the content of the instrument adequately represent the content population of the property being measured? (Nachmias and Nachmias, 1993). The underlying assumption of sampling validity is that every variable has a content population consisting of a number of items (for instance indicators and sub indicators in the Internet mystery-shopping survey instrument) and that a highly valid instrument constitutes a representative sample of these items (Jorgensen, 1989). As of January 1, 2000, there were only fifteen Internet banking service providers in the UK. Because all fifteen Internet banking services are evaluated in this research the question of representatives of the population does not arise. An additional benefit of testing for sampling validity is that it serves an important function: it requires familiarity with all the items of the content population (Fink and Kosecoff, 1985).

Empirical validity

The concern of empirical validity is with the relations between a measuring instrument and the measured outcomes (Nachmias and Nachmias, 1993). It is assumed that if a measuring instrument is valid, there should be strong relations between the results produced by the instrument and other variables. Evidence to support the existence of a relation is obtained by measures of correlation appropriate to the level of measurement. Of the various tests designed to evaluate empirical validity, predictive validity is the most widely used.

Predictive validity is estimated by a prediction to an external measure referred to as a criterion and by checking a measuring instrument against some outcome (Jorgensen, 1989; Nachmias and Nachmias, 1993). In other words, predictive validity is the coefficient between results of a given measurement and an external criterion. The process by which the predictive validity of the mystery-shopping instrument (I) is illustrated in Figure (6.1, next page).

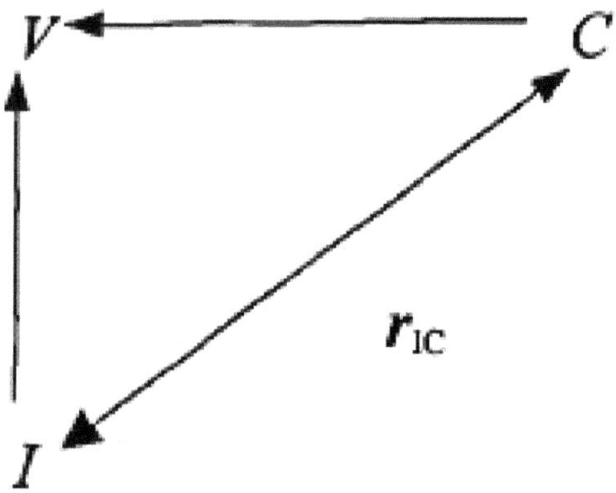

Figure (6.1): Evaluating predictive validity

A variable *(V)* is measured by the measuring instrument *(I)*, and the predictive validity of the instrument needs to be tested. To achieve this, a criterion (C) whose validity is agreed on (say the audited number of customers signed up with each bank) is used. The measurements obtained by *I* are correlated with the measurements obtained by C. The size of the validity coefficient (r_{IC}) measures the predictive validity of the instrument.

Two general issues are to be considered when using the predicative validity test. One relates to the validity of the criterion, and the other concerns the reasons for using a measuring instrument instead of the criterion itself (Nachmias and Nachmias, 1993); for example why not measure customer take up figures directly? In this case, audited customer take-up figures are not available at present and for reasons stated earlier numbers appearing in the popular press are of dubious credibility.

Construct Validity

A construct is an abstraction. It consists in part of "sets of propositions about its relationship to other variables - other constructs or directly observable behaviour (Selltiz, *et al.*, 1976). Theoretical expectations about the variable being measured lead the investigator to postulate various kinds of relationships between the particular variable and other specified variables. Therefore, a construct is validated by observing whether predictions made on the basis of those propositions about its relationship to other variables are confirmed when tested. If multiple sources of evidence yield similar results, that shows the construct's convergent validity. If the construct, as measured, can be differentiated from other constructs, it also possesses discriminant validity (Selltiz, *et al.*, 1976).

Cronbach and Meehl's (1955) logical process of construct validation can be adopted here. Firstly, a proposition that the instrument measures a certain property - say Web page characteristics is set forth. Secondly, the proposition is inserted into the present theory, i.e. that Web page characteristics influence consumer purchase decisions. Thirdly, working through the theory other properties that should be related to the instrument and properties that should exhibit no relation to the instrument is predicted, i.e. this is the process of generation of hypothesis as carried out in chapter four. Finally, empirical data collection will either confirm or reject the predicted relations. If the anticipated relationships are found, the instrument is considered valid. If the predictions fail, there are three possibilities: (1) the instrument does not measure the property, i.e. it does not measure Web page characteristics, (2) the theoretical: framework that generated the predictions are flawed, or (3) the research design failed to test the predictions properly. A decision must be as to which of these three conditions has occurred. Such a decision is based on a careful reconsideration of each of the four steps constituting the validation process.

In view of the distinction among the three types of validity, which test should be used when evaluating the validity of the measuring instrument? Nachmias and Nachmias (1993) assert that there is no simple solution. They quote a group of experts from different disciplines to recommend that through examination of a measuring instrument include information about all three types of validity. Therefore, in the fust phase of the construction of the measure, theories were evaluated that would serve as the foundation for the instrument, this was achieved through chapters two to four. Next, a content population of items from which a representative sample is to be drawn was defined, and in this research the problem of representativeness does not arise since the entire population is evaluated. And finally, correlating it with an external criterion will assess the predictive validity of the instrument.

6.6.3.3 Other Criteria Governing a Good Research

How good the research design is, can be determined to a great extent by the criteria discussed above. However, those criteria alone do not ensure the quality of the study itself and the results since they relate mainly to the research design and not the actual implementation. The criteria below are commonly used as prerequisites for completion of good research and therefore will be crucial for the success of the study.

Reliability and Generalizability

Reliability is a central concern because measuring instruments are rarely completely valid (Nachmias and Nachmias, 1993). Reliability refers to the extent to which a measuring instrument contains variable errors, that is, errors that differed from observation to observation during any one measuring instance or that varied *from time to* time for a given unit of analysis measured twice or more by the same instrument (Nachmias and Nachmias, 1993). Therefore, the objective is to be sure that, if a later investigator followed exactly the same procedures as described by an earlier investigator and conducted the same study all over again, the later investigator should arrive at the same findings and conclusions (Yin, 1994). The goal of reliability is to minimise the errors and biases in a study.

Cronbach, *et al.,* (1963) introduced a revision to the traditional concept of reliability.

They maintain that the chief concern of reliability theory is to answer the question

> *"to what universe ofpotential measurements do with wish to generalise?"*

(Cronbach, *et al.,* 1963, p. 147).

Thus instead of reliability, the idea of generalizability implies the following; (1) what one really wants to know about a set of measurements is to what extent and with respect to what properties are they like other sets of measurements that one might have taken from a given universe of potential measurements? And (2) to what extent and with respect to what properties do they differ from other measurements that one might have drawn from that universe of potential measurements? Therefore this means that the likeness and difference questions of potential measurements, is a function of the limits of generalizability of the results based on the set of measurements. By extension this means that a particular relation among measurements to be the evidence of reliability or generalizability depends on how likeness, difference of conditions and measures are defined.

Several measures have been taken to ensure reliability and generalizability in this study. All Web pages have been downloaded, and stored for retrieval at a latter date. Careful records have been made of the process of accessing Web sites and of the downloading process. Hence the potential for errors from one observation to another are minimised. Therefore, if the study is replicated at a latter date, and the investigator follows exactly the same procedures as described in this research and conducts the same study all over again, the investigator should arrive at the same findings and conclusions reported in this study.

Relevance of Findings

Evered and Louis (1981) find a quantitative (objective) approach methodologically precise but often irrelevant to the reality. On the other hand, they explain how in the qualitative research approach, features are noticed and identified through an interpretative, iterative process whereby data and categories emerge simultaneously with successive experiences. They give this as the key contributor to the richness of the situation specific understanding gained. This discussion illustrates the significance of making a research based on an objective approach, relevant to the world. However, this criticism of quantitative research is mainly aimed at studies conducted in isolation, possibly using simulation techniques, and is therefore of limited relevance. It is possible to argue that this study is also based on simulation, however as already discussed in this chapter simulation in the case of Internet banking services simulation equates to actual service delivery. Additionally, the study is not made in isolation.

6.7 Chapter Summary

This chapter began by illustrating the background of the methodology problem. It identifies the significant differences in the nature of Internet banking services compared to traditional banking services which have implications for the methodology. In particular, it is seen that the need for attitudinal research is reduced. Reviews of previous research methodologies in financial services research show that these are predominantly attitudinal research methodologies. It is argued that there is a case for an alternative research methodology for this research in light on the unique qualities of Internet banking services and the opportunities it presents. A new research methodology, mystery-shopping on the Internet, is developed, using the best elements of traditional mystery-shopping methodology. Evaluation of the proposed mystery-shopping research methodology finds that there are considerable advantages in using it in Internet banking services. Moreover it illustrates that these advantages are applicable for both the academic and the commercial world. The discussion is extended to evaluate how the proposed research methodology measures up to principles of good research design. Overall, this chapter has contributed to this research work by adopting preceding best practise to generate a sound and novel research methodology.

Chapter 7.

The Research Process

7.1 Introduction

Previous chapters have been important in developing the research framework and methodology for this thesis. This chapter operationalises the mystery-shopping research methodology presented in the sixth chapter. It achieves this by describing the development of survey instruments and conceptualising key variables. During this process it is important to make cross-references to the working hypotheses so that the validity and the appropriateness of the research instrument is maintained. The chapter is divided into three sections.

The first section explains the procedures adopted to develop the measurement instrument, i.e. the Internet survey scorecards. In particular the sources of information used in developing the Internet scorecards are examined. The second section explains in detail the measures of each of the Internet scorecards which will be used for the mystery-shopping survey, this section is divided into four parts, each dealing with separate but interrelated scorecard series. Finally, the different Internet banking services that are subjected to the mystery-shopping survey are introduced. The chapter concludes with a summary.

7.2 The Development of Measurement Instruments

The task at hand is to develop an instrument that can evaluate the hypothesis by generating adequate measures. As explained in the previous chapter, the mystery-shopping methodology applied to Internet banking services is evaluated on four interrelated levels,

- Web site evaluation,
- demonstration component evaluation,
- email/query application evaluation and
- overall evaluation.

To comprehensively evaluate each of these levels, a series of scorecards, consisting of both subjective and objective measurable indicators are developed. Taken as a whole, these capture adequate measures required for the testing of hypotheses. In the previous chapter the problems associated with ordinal scales, which have been used in previous studies in the evaluation of subjective measures, were discussed. In order to mitigate these weaknesses, a bi-polar semantic scale was suggested. The scale, 10cm long, is calibrated at one-centimetre intervals, and provides a point of reference for researchers from which they evaluate their subjective assessment of measures. Once the assessment is complete, the data were then fed into a computer database for analysis.

Previous chapters have been instrumental in drawing out gaps in the literature, highlighting areas where more work and a better understanding is required. Each of these gaps or areas gives rise to hypotheses that encapsulate the central research question. The measures in the scorecards are designed to test the hypotheses. Therefore the literature review constitutes the foundation for the development of scorecard measures. Two other sources of information, namely, Internet discussion forums and online banking surveys, were used to strengthen the scorecard measures. The following two sections will discuss each of these sources of information in tum. Each section will pay particular attention in describing the characteristics of the source information, the purpose they serve and the validity of using them in academic research.

7.2.1 Internet Discussion Forums

An Internet discussion forum represents a heretofore unavailable medium of expression and dissemination of information and opinion by individuals or groups of individuals. A discussion forum is a particular Web site, or a portion of a Web site, with the facility for individuals to express their opinions on a certain topic or a series of related topics. Some authors refer to such forums as online-communities, others may refer such areas on the Web that they contribute as bulletin-boards or discussion-boards. For example, if the forum discussion is on "Internet Banking", by and large all contributors limit their contributions to this broad subject and related areas. Forums may be, (1) free-standing, (2) promoted by an individual or group of individuals out of common interest, or (3) they may be hosted by a particular organisation on their Web site as a means of building a Web site with high usage, thereby achieve high 'hit-rate' from Web surfers. Inevitably if an organisation hosts a forum there must be a clear association of the organisation's activities and forum discussion topic, for example, a financial services portal may host a forum on Internet banking. Some of the advantages accruing to the host organisation Web site by these forums are,

- provides wide publicity and by extension helps to build the brand of the host organisation,
- a high-usage Web site attracts Web banners and other forms advertisements that provide a stream of revenue, and
- it helps to build a sense of community among users of the site, and these are likely to be frequent users of the site. These users are a captive audience for any goods or services that the host organisation might wish to sell.

For the users, it provides a means for them to express their views, achieve a wide audience and provide a feeling of being belonging to a community of individuals that share the same interests.

There are two main methods of contributing to these forums. On one hand, an individual can commence a new topic, this is called starting a new "string", or they can respond to a string that is already present on the forum. When a contribution is made, most forums record the date and time of the contribution, the name or pseudonym of the contributor. It is also the case that, many of these organisations that host forums, require contributors to become a member of that organisation by filling an online application form which builds a personal profile of that individual. Usually no monetary consideration is charged to become a member. By acquiring members in this method, organisations are building up a valuable database of personal profiles which can be used for marketing purposes, which is an additional benefit.

7.2.1.1 Forum Contributor Characteristics

The characteristics of discussion forum contributors can be summarised as follows.

1. The contributors represent a vociferous self-selecting sample from the millions of Internet users.
2. They are fairly conversant with the workings of the Internet in general, and how to contribute to discussion forums in particular.
3. Most contributors are already using Internet banking services. These individuals are well versed with the workings of Internet banking services, and possess an above average degree of hardware and software know-how. On the other hand if they are not already using Internet banking services, they are currently considering the use of Internet banking services and requires impartial information from current users.

4. While the majority of contributions to the forum deal with an unpleasant encounter or a series of encounters, there are also a significant number of contributors with narrations of satisfactory encounters. This is helpful in developing measures that would capture such aspects on a service.
5. The contributors to these forums comprise of like-minded individuals. Their observations are, in most cases, straightforward: it is easy tell whether one Internet banking service is better than another, and if others disagree, they will express their opinion. False and unsubstantiated information can spread fast, but it can be quickly debunked. Therefore it is difficult to manipulate opinion among forum contributors.

Consumers intent on purchasing Internet banking services or those who are already consuming Internet banking services are themselves a self-selecting group of individuals from the wider population. Out of this group, a further group of self-selecting individuals decide to contribute to a forum of this nature.

A question that needs to be addressed at this point is to what extent is such a discussion forum an acceptable source of information in academic research? More specifically what purpose will a forum fulfil in this research?

7.2.1.2 *Purpose*

As stated earlier, a forum is used primarily to aid in the generation and refinement of measures. It is important to stress that it is not used as a survey, to collect primary data. The suggestions and comments made by the contributors help this process by highlighting issues that may escape the researcher. In effect it shares some of the characteristics of a focus group. Additionally, it may act as a tool for triangulation, i.e. if the experiences of contributors echo the findings of this research, it also helps to verify research findings.

Next it is prudent to consider whether is it possible that the characteristics of a forum pose a threat to the reliability and the generalizability of the research findings?

7.2.L.3 *The Validity of using Forums in Academic Research*

The goal of reliability is to minimise the errors and biases in the study. By extracting information from a forum with the aforementioned characteristics the study's reliability can be verified.

Two discussion forums are used for this analysis. The first forum, the "Online Banking Discussion" is available at [http://www.iii.co.uk/online-banking/?type=discussion]. It is operated as a part of the 'interactive investor international' "iii" found at [http://www.iii.co.uk/]. Although the name of the parent organisation suggests an "international" dimension, the discussion is limited to matters concerning Internet banking services available to UK residents and nationals. Of additional value is the contribution made by UK nationals working in other countries as expatriates, operating their UK accounts over the Internet. This provides an insight to the operation of bank accounts over national boarders. This is one of the unique characteristics of Internet banking services that cannot be matched by terrestrial banking. The contributor content is not moderated by iii, however inappropriate use of language, for example obscene language, is censored. For the purposes of analysis, 257 contributions were downloaded into an Excel database. In addition to the forum content, the name of the contributor and the date that contribution is recorded. The second discussion forum is found at 'The Motley Fool UK' [http://www.fool.co.uk]. This is a Web site that has many discussion forums, and 489 contributions (consisting of contributions up to 01/01/2000) from the particular discussion forum relating to Internet banking services were analysed.

Does the forum satisfy sampling validity criteria? To reiterate, the discussion forum is not used as a survey, rather as a tool to assist in the process of generating scale measures. Therefore sampling validity of discussion forum contributors is not a threat to the robustness of the research.

It was earlier stated that research methodology has to fulfil the validity criterion in order to reduce measurement errors. The mystery-shopping methodology was deemed to fulfil face validity criteria because the scale measures were developed after an extensive literature review. The extraction of data and information from the forum augments this process by adding to and refining the scale measures.

Perhaps the most significant contribution from the forum analysis is the provision of empirical validity to the mystery-shopping methodology. Jorgensen (1989) identified that predictive validity can be estimated by a prediction to an external measure referred to as a criterion and by checking a measuring instrument against some outcome. In this instance, a forum contributor's narration of their experience can be construed as the criterion whose validity is agreed. The measurements obtained from the scorecards can be correlated with the data obtained from the forum to measure the predictive validity of the instrument.

7.2.2 Online Surveys and Consultancy Reports

As previously stated in section (4.4), one of the attributes of the Internet is the ability to collect and disseminate information cheaply, efficiently and quickly. These attributes have given rise to a large number of online surveys being conducted by various organisations. These surveys have a variety of characteristics, including;

1. Like Internet discussion forums, respondents are a self-selecting group.
2. One of the primary objectives of surveys is to attract media attention (except perhaps for those conducted by academic institutions), thereby increasing the profile and the visibility of the organisation conducting the survey. Another equally important objective is to increase the number of visitors to the Web site and increase Web traffic, since people will come frequently to a Web site that contains valuable information for free.
3. Survey reports could be a public relations exercise and it is doubtful whether the questions to which the respondents are asked to respond are articulated after extensive literature reviews, as it is the case in academic studies. The surveys are more likely to contain questions that attract the attention of the Web surfer.
4. Consequently, the surveys deal with macro issues and are unlikely to pay attention to micro issues with very detailed questions.
5. Web surveys need to be kept to a reasonable length for Web surfers to respond to them quickly, and this denies the opportunity of inserting a large number of detailed questions. This means that the likelihood of the respondents misunderstanding the questions is higher and consequently the results of such surveys need to be viewed with caution.

While there are a large number of online surveys on the Internet, for the purposes of this thesis, survey reports from the following Web-based organisations are referred to.

1. 4anything.com
2. Banking.Com
3. Gomez.com
4. GVU
5. IFBG Gottingen
6. iii (Interactive Investor International)
7. UK Internet Banking.com

8. ZD Net UK

The first two survey studies are conducted by academic organisations and therefore may lack commercial bias that may be found in the rest.

Apart from the above type of survey reports, consultancy reports have been another source of information. These consultancy reports are complied by management consultancy companies, and contain surveys of one form or another as an integral part of the reports. The reports are either compiled for a specific client requirement, or as a general report to be sold to a wide range of companies or individuals that may require such reports for investment or other purposes.

7.2.2.1 Purpose

On balance, in spite of the recognised disadvantages, it is still worthwhile to analyse these commercial surveys and consultancy reports. The use of these surveys and consultancy reports is to supplement the knowledge gathered by the literature survey and forums, so that scorecard measures can be refined. It is reiterated that reports are not used as a survey to collect primary data.

7.2.2.2 The Validity of Using Online Surveys and Consultancy Reports in Academic Research

Characteristics of consultancy reports need to be kept in perspective, and an informed judgement is required in their use of in academic research. In particular the following salient features of these surveys and consultancy reports are noted.

Firstly, the reports are produced as a method of generating revenue by the authors. Therefore the findings may be biased towards the eventual buyers or sponsors of reports. Secondly, the statistical rigour in the surveys is questionable. In a research project commissioned by the office of the e-Envoy and the Cabinet Office, where the author was a team member, it was found that the data collection procedures in a large number of these consultancy reports are not statically rigorous and do not measure up to good statistical practise. Reasons for this apparent lack of statistical rigour could be the type of audience for which these reports are intended and perhaps most importantly resource constraints. Thirdly, the reports appear to concentrate disproportionably on forecasting the future based on very little empirical evidence, but built primarily on the 'expertise' of the report authors. Finally, the circulation of these reports are very limited, hence a report tends to be very expensive, typically several thousand pounds. An implication of this is that organisations that buy the reports may be favourably treated in the report. Nevertheless, as already stated earlier, in the absence of superior secondary sources of information, consultancy reports from the following consultancy companies were referred to in the development of the research instrument scale measures.

1. Datamonitor
2. Forrester
3. Giga Information group
4. Jupiter
5. Lehman Brothers
6. Warburg Dillion Read

Now that information sources for the development of Internet scorecard measures have been discussed, it is appropriate to describe the scorecards in detail. In the previous chapter the need to develop a four levels of interrelated scorecards were presented. Therefore the next section will introduce each of those four levels of scorecards. In addition an indication will be made as to the relationship between a particular scorecard and the hypotheses that it will test. The section begins with practical and procedural aspects of carrying out a mystery-shopping scorecard survey of Internet banking services.

7.3 Mystery-Shopping Surveys

Due to the ephemeral nature of Web pages it is essential that a copy of each page is stored for analysis. Therefore, each Web page is downloaded to a computer hard disk. These were then transferred to a recordable compact disk as a precautionary measure and for the added versatility of being able to retrieve the information on any computer with appropriate hardware.

Internet Explorer (IE) version 5 was used for browsing and downloading. This was because, in comparison to the second most popular browser, Netscape, only IE is capable of downloading all the graphics on a page along with the text, without having to individually download the graphics (which are separate files). IE achieves this process by creating a separate directory for each page downloaded. Web pages are authored either in a single language, i.e. Hypertext Markup Language (HTML) or Java, or as is increasingly the case, in a combination of HTML and Java. The Web pages deliver, text in various positions and contexts, graphics (both still and moving images), colours, sounds, etc. on to the consumer's computer terminal. During the saving process, at times it was necessary to change the default page title so that the number of characters did not exceed the limitations on the operating system. This process ensured that all pages authored in HTML in its entirety were downloaded to the computer. However the same cannot be said for Web pages, in their entirety or portions, authored *in* the *Java* language. Java is a language that is being used in an increasingly widespread manner for it has several advantages. Some of these advantages are,

- It is an object oriented language, therefore Java based pages and applications can run on a wide range of apparatus, including WAP phones, TVs, computers, etc., so long as they support Java
- Java based applications consume less memory for execution
- Java based applications can be more robust in their execution in comparison to HTML
- Java is more suited to interactive type of applications
- Java is more suitable for animation, drop down menus, and other similar features.

Due to these advantages, among others, Java is being increasingly used in Internet banking Web sites. Most of these Java based Web pages are designed to communicate constantly with the host and the user, thereby making these Web applications unstable or non-operative in offline mode. Therefore some of these pages could not be saved, and even if they were saved they are not retrievable. This is a significant shortcoming for the analysis, and threatens replicability. There is no solution to this problem.

As explained above, mystery-shopping surveys of Internet banking services are carried out using four interrelated Internet scorecards, each scorecard comprising of several sections. In presenting these scorecards, the discussion is arranged as follows. A general overview of the scale measures for each of these sections is discussed.

The functions of each of the four scorecards are explained under sub headings. Since each of the scorecards attempt to test a single hypothesis or a number of hypotheses, such explanations are made with references to the appropriate hypothesis or hypotheses.

Many of the scale measures were subjective, however, where possible reviews by independent sources were also utilised as additional support for the subjective scale measures. The evaluation of some of the quantitative measures was augmented with a third-party software programme ('Alexa monitor') that generated measurements. This piece of software is freely available, and is downloadable from [http://www.alexa.com/]. The Alexa monitor is a Web navigation service that works the browser and accompanies user as they surf, providing useful objective information about the sites that they are viewing and suggesting related sites.

7.3.1 The Web Site Mystery-Shopping Evaluation

The mystery-shopping scorecard enables the evaluation of how all attributes of the Web site contribute to the consumer purchase intentions of Internet banking services by firstly aiding in the information search and acquisition, and in evaluating the service quality. The scorecards were subdivided under two titles, 'Web site content' and 'Web site characteristics'. The Web site mystery-shopping section contributes to the measurement of three hypotheses, namely, *H2, H8* and *H9.*

7.3.1.1 Website Content

This section presents the measures that assist in the testing of,

H 2: Purchase intentions, mediated by perceived information search quality, are influenced by the contents of an Internet banking service Web site.

Fourteen measures are utilised to measure the information and content on the Web sites, and Table (7.1, next page) shows a detailed description of the measures and the theoretical support for selecting specific measures. It must be noted that the references on the theoretical support column, on Table (7.1) and on other tables that follow in this chapter, are by no means and exhaustive list, but provides an indicative list of references.

	Title/Measure	Explanation	Theoretical Support
1	Clear and concise text	The content is un-ambiguous and brief	Gehrke and Turban, 1999; Bell and Tang, 1998; Gardyne (1998); Heath (1998)
2	Relevancy	The content is useful and the information contained is relevant.	Davis, 1989; Doll and Torkzadeh, 1988; Muylle, *et al.*, 1999, Bell and Tang, 1998; Gardyne (1998)
3	Accuracy	The content is accurate.	Doll and Torkzadeh, 1988; Muylle, *et al.*, 1999; Gardyne (1998); Blower (1998)
4	Comprehensibility	The content is easy to understand.	Doll and Torkzadeh, 1988; Muylle, *et al.*, 1999; Gardyne (1998)

5	Comprehensive-ness	The Web site conveys all information necessary. The need for clarifications is minimised.	Doll and Torkzadeh, 1988; Muylle, *et al.*, 1999; Gardyne (1998)
6	Amount of site content	There is a balance between number of pages and amount of site information.	Doll, et al., 1994; Muylle, *et al.*, 1999, Bell and Tang, 1998; Gardyne (1998)
7	FAQs	The 'Frequently Asked Questions (FAQ) section is comprehensive. In instances where there is no dedicated FAQ section on the Web site, the objective is to assess how comprehensively the site content can be used to quickly and completely answer consumer's questions.	Gehrke and Turban, 1999; Sterne, 1999; Janal, 2000
8	Ease of under-standing	The language used is easy to understand. The amount of legal jargon is kept to a minimum. A higher score indicates easy to understand and information is presented in a concise manner, not taking long to read and understand.	Schillervaert, *et al.*, 1998; Sterne and Priore, 2000; Sterne, 1999

9	Disclaimers	The amount of disclaimers is kept to a minimum. The site is very clear and to the point on what are the responsibilities of each party and how they are executed.	Johnston, 1997; Sterne and Priore, 2000; Sterne, 1999
10	Format	The format adopted for the presentation of information is suitable. This encompasses the arrangement of paragraphs, width occupied by text, scrolling down/sideways, bullet points, numbering of points, etc.	Jayawardhena and Foley, 2000
11	Up to date	The information found in the Web pages is most up to date. There is no conflicting content indicating updating inconsistencies within the site.	Gehrke and Turban, 1999; Sterne, 1999; Chaffey, *et al.*, 2000
12	Amount of downloaded content (no. of pages)	The number of 'relevant' pages downloaded. This reflects the number of pages a consumer may have to consult to get information about the banking service.	Sterne and Priore, 2000; Strauss and Frost, 1999; Janal, 2000

13	Total number of pages on Web site - (Alexa count)	"Pages" is the number of Web pages on the site. Page data is based on Alexa's periodic Internet crawls. Page data does not include pages excluded by robots.txt, cgi-bin, and other restrictions.	Sterne and Priore, 2000; Strauss and Frost, 1999; Janal, 2000
14	Freshness (Alexa monitor)	Freshness is an indication of how often a site is updated. Rating is determined by analysing the modification dates of pages found during Alexa's periodic Web crawls. Freshness is the average age of the top pages of a site (not including cgi-generated pgs.)	Gehrke and Turban, 1999; Sterne, 1999; Chaffey, *et al.*, 2000

Table (7.1): Content measures

The multimedia content characteristics are expected to influence the perception of information relevancy, accuracy, comprehensibility, and comprehensiveness, as well as the perceived language customisation and layout (Muylle, *et al.*, 1998; Doll and Torkzadeh, 1998). This is by far the most prescribed recommendation in the literature. Gardyne (1998) provides nine recommendations: brief sentences, bulleted lists, short pages, highlighted keywords, colourful, descriptive paragraph headings, one idea per paragraph, most important point first in each paragraph, and half the word count (or less) of conventional writing. Blower (1998) adds proper spelling and grammar to this list. Providing contact information and installing 'mail to' links on every page aids in the retention of consumer interest and provides feedback (Smith, 1997). Using simple background colours and textures, which are pleasant and easy on the eye enhances the content appeal (Heath, 1998). Information accuracy is mainly related to the timeliness of the Web site content. Regular updates of changed information (for example, interest rates) are important. According to (Gehrke and Turban, 1999), providing a FAQ (Frequently Asked Questions) section and regular changes to site is an worthwhile method of keeping information accurate and up to date.

The scale measures are designed so that they evaluate the degree to which a consumer may be able to gain an understanding of the Internet banking service (Schillervaert, *et al.*, 1998). Also of importance is to recognise and evaluate the dynamic nature of the Internet in terms of up to date content and record updating inconsistencies (Gehrke and Turban, 1999). Out of the fourteen measures, eleven measures are subjective, where the evaluation was carried out on the bi-polar semantic scale introduced earlier. The remaining three measures are objective.

7.3.1.2 Web Site Characteristics

This section presents the measures that assist in the testing of,

H 8: Purchase intentions, mediated by perceived Internet service quality, are influenced by Website characteristics.

In order to influence purchase intentions, such information needs *to* be (1) comprehensive and cover all aspects of the Internet banking service, (2) presented in a manner suitable to the Internet, and (3) accessed easily using methods that are suited to the Internet. Table (7.2, next page) shows a detailed description of the measures that would evaluate the aforementioned aspects.

	Title/Measure	Explanation	Theoretical Support
Ease of use/Navigation			
1	Entry guidance	It is easy to arrive at the site from the wider Web	Muylle, *et al.*, 1999; Bell and Tang, 1998
2	Number of links from other Web sites (Alexa count)	The number of Links In is an indication of reputability. "Links In" is the number of different Web pages that link to this particular Web site.	Sterne, 1999; Chaffey, *et al.*, 2000
3	Ease of use	The website is perceived to be friendly	Doll and Torkzadeh, 1988; Muylle, *et al.*, 1999
4	Web site structure	User perception of the manner in which different parts of the website are linked up	Muylle, *et al.*, 1999; Bell and Tang, 1998

5	Navigation consistency/site index	Navigation within the website is consistent. It is easy for a user to access/use the index. The index is effective tool.	Gehrke and Turban, 1999; Sterne, 1999
6	Usefulness of hyperlinks	How many and how useful are the hyperlinks? Hyperlinks are accurate lead to the intended location.	Gehrke and Turban, 1999; Schillervaert, *et al.*, 1998; Haine, 1998
7	Hyperlink connotations	How the user interprets hyperlinks	Muylle, *et al.*, 1999; Haine, 1998; Sterne 1999
8	Hyperlink encoding	How easy is it to for users understand what is meant by hvoer-links *(and* the contents therein)	Muylle, et al., 1999; Haine, 1998; Sterne 1999; Chaffey, *et al.*, 2000
9	Search engine/facility	Efficiency and effectiveness of the search facility.	Gehrke and Turban, 1999; Schillervaert, *et al.*, 1998;
10	Homepage hyperlinks	There is a clear route to the home page on every website page and it is prominently displayed.	Gehrke and Turban, 1999; Schillervaert, *et al.*, 1998;
11	New browser window	On clicking certain links, a new browser window opens	Sterne, 1999
Design			
1	Layout	The Web site layout is appropriate.	Doll, *et al.*, 1994; Muylle, *et al.*, 1999; Bell and Tang, 1998;

2	Intelligent use of frames?	The site uses frames. Low reliance is indicated with a high score. Heavy reliance leads to distorted pictures with different screen settings, downloading can take longer, more prone to errors if the Internet connection is not stable/fast, makes book marking difficult.	Gehrke and Turban, 1999; Bauer and Scharl, 2000; Brernser, 1997
3	Intelligent use of tables	Text arranged in tables/column format reducing the need for frames. It is also a useful feature to incorporate so that different screen settings can be accommodated.	Schillervaert, *et al.*, 1998; Sterne, 1999
4	Use of colours	Choice of colours is easy on the eye, attractive and is appropriate.	Schillervaert, *et al.*, 1998; Bauer and Scharl, 2000;
5	Use of graphics	Web site contains simple and meaningful graphics that are appropriate.	Bell and Tang, 1998; Gehrke and Turban, 1999
6	Likelihood of distortions	Differences in screen resolutions lead to distortions. Very thin lines and icons could make the user get confused, or in worse case scenarios the user may not see the icon at all. A higher score translates to little distortions.	Ho, 1997; Sterne, 1999
Others			

1	Online help features	The availability of online help features or similar features and it is useful.	Sterne 1999 and 1996; Chaffey, *et al.*, 2000
2	Interactivity	(I) Interactive features that allows users to customise preferences. (2) Interactive features like online calculators, etc. are available on the Web site. (A breakdown of these features are covered functionality)	Doll, *et al.*, 1994; Muylle, *et al.*, 1999
3	Language customisation	The language can be tailored to the prospective user.	Muylle, *et al.*, 1999
4	Browser suitability function	A system to check the suitability of the browser that the user is presently using.	Janal, 2000; Sterne, 1999

Table (7.2): Webpage characteristics

Once a consumer enters a bank Web site, they begin to look for information within the site. To achieve their objectives consumers may use a combination of hyperlinks, navigational tools, an index, search engine, opening of new browser windows, etc. (Gehrke and Turban, 1999; Schillervaert, *et al.*, 1998; Haine, 1998). There two primary uses of a new browser window opens when one clicks on a hyperlink.

1. It is a security feature. A new copy of the client (i.e. consumer's) browser window will only accept and send coded messages to and from the Bank Internet server.
2. The user does not get 'lost' within the site. In the alternative, the site configured in such a way the need to open new windows is minimised/eliminated.

Hyperlinks are an essential part of the Web, hence it is imperative that hyperlinks are designed carefully so that consumer connotation (how the user interprets hyperlinks) is not misrepresented (Muylle, *et al.*, 1999).

The hypertext of the multimedia content also influences entry guidance, Web site structure, and the navigational efficiency. Without efficient and user-friendly navigation, the user is likely to get confused, lost, or: frustrated and site for good. In this regard, eleven measures evaluate the processes a consumer goes through from the point of entry to the Web site and navigating within it (Bell and Tang, 1998).

To begin with the use of well labelled and accurate (no broken) links are of paramount importance (Haine, 1998). Navigation consistency is proving to be a debatable topic. Wilson (1998) favours the provision of a number of ways to navigate within the site. Berst (1998) disagrees, by favouring a consistent approach throughout the Web site. The length of a Web page is also an area where there seems to be some contention among the authors. While Janal (2000) favours long pages with links within the page, Heath (1998) disagrees by arguing that long pages are confusing because users scroll down to read a paragraph, then return to the top of the page only to select the same paragraph again from the index. Gehrke and Turban (1999), recommend that a large site with several underlying pages should have a map or guide that allows the user to jump to any page from any page. Many authors including Tadger (1998) support the provision of a navigation bar on each page. The provision of an effective search engine aids in the navigation within the site. However this is more applicable to large sites containing a large amount of pages or content, or for sites with a lot of randomly accessed content (Haine, 1998).

As a general rule, many authors agree that search engines should be prominent and easy to use. Distinctive 'hot' buttons also aid in the navigation process, and have a personalised effect on the Web page. Some authors (Busch, 1997) suggest the usage of unusual shapes and colours, in addition to the traditional rectangular or circular shapes. Rollover effects for buttons, whereby the movement of the cursor over the button pops open a text box describing in more detail how the button works, enhance navigation efficiency (Kramer, 1997). Gehrke and Turban (1999) observes that a large number of authors recommend avoiding the usage of frames. One of the many reasons for using frames is when a Web site designer wants to make the common features, links, logos, etc. visible from all underlying pages irrespective of the place a user might be reading. But, Tadjer (1998) claims that frames have been overused in many Websites. Many authors (Jayawardhena and Foley, 2000; Tadjer, 1998; Wilson, 1998; Berst, 1998) have commented on the disadvantages of frames. Some of these disadvantages are.

1. Frames make book marking difficult
2. Some browsers do not support frames
3. Frames divide the screen into windows that necessitates additional

scrolling movements (both vertically and horizontally) depending on the screen resolution

4. Frames slows down page loading
5. Frames complicates printing process on some browsers
6. Often content within the frames are skipped by search engine 'spiders'
7. Frames are resented when site owners use them to link to from external sites
8. Some frames do not scroll down when they should not and some frames produce an unwarranted patchwork effect.

Bremser (1997) supports the use of frames only if the frame layout is made simple and all given control to the user, i.e. allow the user to toggle the frames on or off.

Web design features are evaluated for their impact on purchase intentions. In addition to the obvious aspects of colours, graphics and layout, aspects such as use of frames and tables, likelihood of distortions under a multitude of screen specifications and settings, etc. are evaluated (Bauer and Scharl, 2000; Schillervaert, *et al.*, 1998;). And finally, other aspects that contribute to Web page characteristics, such as interactivity of the Web pages or sections of them, browser suitability functions, are evaluated for they are hypothesised to have an impact on purchase intentions (Doll, *et al.*, 1994).

7.3.1.3 Performance of the Web Site

Performance of an Internet banking service Web site is evaluated using twelve measures, and these appear on Table (7.3, next page) and assist in the testing of,

> ***H 9: Purchase intentions, mediated by perceived Internet service quality, are influenced by the performance of the Website and related components.***

These measures are divided into three groups. Broadly speaking, operation of the Web site encompasses all the requirements that the consumer and their computer have to comply to operate or access the Web site. This includes pre-requisites of software requirements, ways in which software have to be configured, particular ways in which the browsing process should be undertaken and conditions under which information is sent over the Internet (Gehrke and Turban, 1999; Bauer and Scharl, 2000). Thereafter, the second group of measures evaluate aspects relating to gaining access to the Web site (Bell and Tang, 1998; Bauer and Scharl, 2000).

	Title/Measure	Explanation	Theoretical Support
Operation			
1	Animation/multimedia plug in requirements	Plug-in software requirements leads to slow page loading, and also slows/prevents user from scrolling down/moving within the website.	Gehrke and Turban, 1999; Haine, 1998.
2	Any additional software to download?	Additional software (such as java applets) downloading is necessary for the operation. These are bespoke software.	Sterne, 1999; Bauer and Scharl, 2000; Janal, 2000
3	If so, uninstall facility?	If downloads are needed, can these be uninstalled?	Sterne, 1999; Bauer and Scharl, 2000; Janal, 2000
4	Information retention in computer	Some information is retained in the computer (account related), and can be accessed later.	Strauss and Frost, 1999, Chaffey, *et al.*, 2000
5	Offline operation	Synchronised browsing. It is possible to operate the Web site when there is no connection to the Internet.	Bauer and Scharl, 2000;
6	Cookies	The Web site store information in a small text file, called a "cookie," on the hard disk.	Bauer and Scharl, 2000; Strauss and Frost, 1999; Chaffey *et al.*, 2000
Access to Web site/operation			
1	Web congestion	The level of Web site congestion. How well is the site is Bauer able to handle traffic?	Bell and Tang, 1998; and Scharl, 2000;

2	"Bugs" in the operation	The presence of unstable Web pages or other problems with the site. Require repeated login in (reloading) to achieve a stable connection.	Bauer and Scharl, 2000;
3	Planned maintenance of website (hrs per week)	The number of hours the banking service unavailable for customers	Bauer and Scharl, 2000; Janal, 2000
Web site speed			
1	Downloading time	Time taken for downloading the Web pages. A comparative measure is taken here since a more objective measure is taken below in (3).	Muylle, *et al.*, 1999; Bauer and Scharl, 2000; Hamilton, 1997; Wilson, 1998.
2	Offline operation	Time taken to open a Web page from files resident in the computer in offline mode. Also includes time required for executing offline software, if required.	Bauer and Scharl, 2000; Janal, 2000
3	Access speed (in kb/s)	The Speed rating indicates a site's effective transfer rate measured in kb/s. It is measured during Alexa's periodic crawls of the Internet.	Sterne, 1996; Sterne and Priore, 2000.

Table (7.3): Performance of the Web site

The hypertext connection of the multimedia content influences the Web site speed, or the rate of page loading (Muylle, *et al.*, 1999). According to a survey conducted by Hamilton (1997) speed (i.e. slow speed) was the number one complaint of Web users (77%). Later surveys (Gehrke and Turban, 1999) have shown that that the problems associated with Web site speed

(22%) have declined, however it is still the main concern among Web users. Several factors influence the speed of page loading. The primary reason for slow loading is attributed to complex graphics (Wilson, 1998). Wilson (1998) recommends that graphics be kept simple, meaningful and relevant to the context. Animation is a contributor to slow page loading; hence it is prudent to limit the use of animation and/or multimedia plug in requirements (Gehrke and Turban, 1999). The notion that 80/20 rule should apply to Web pages is gaining prominence (Nunley, 1998), i.e. information content should account for 80% of the site. Haine (1998) recommends that if a designer thinks it is necessary to have animation, it should be made to stop cycling after ten seconds to allow the user to scroll it off the screen. Usage of thumbnails, which typically contain only about 1OKB, are an effective way to utilize graphics without significantly increasing the page loading time. Thumbnails give the user a choice of whether or not to wait for the loading of a larger picture without forcing it as a default (Gehrke and Turban, 1999). Providing a 'text-only' choice increases the speed and allows the user of a greater variation of browsers (Heath, 1998). Sklaire (in Janal, 2000) describes this term as allowing text to load first, followed by graphics. This allows the user to read the content while the graphics are loading.

7.3.2 Demonstration Mystery-Shopping Evaluation

The demonstration component on an Internet banking service Web site is useful for two reasons. Firstly, a demo greatly enhances a consumer's comprehension of the Internet banking service experience. Therefore, it aids in the information search and acquisition. Secondly, perhaps more importantly, it provides experience qualities by simulating the purchase and use of an Internet banking service. Therefore it is helpful in giving an indication of the functionality of the account and how an actual account would operate.

Except for one Internet banking service, egg, all others being evaluated for this research incorporated a demonstration component. As noted above in the description of Web pages, the demonstration or 'demo' as these are popularly referred to, can be authored in HTML or Java or in a combination of both. Once again, downloading problems highlighted earlier apply with no suitable solution. Further, in order to view some demos, a specific file (in most cases an 'executable' file) is required to be downloaded into the computer. Once the file is saved in the hard disk, it is 'run' (executed) and the demo will commence. The majority of demos work as an integral part of the bank Web site and does not contain an executable file to download. Further, some demos require the consumer's computer to have additional software, such as shockwave, animation software, apart from the standard browser software for the consumer to observe the range of available functions in the demo.

Many of the Internet banking service Web sites in combination with the demos play a crucial part in attempting to deliver a potential customer a 'true feel' for the actual operation of an account, by simulating the functionality of the accounts. The demonstration component also allows a consumer to experience some of the service quality aspects. This is achieved by linking different sections and components of the account, such as communication methods, functionality, etc. Therefore, the demonstration and the account functionality mystery-shopping evaluation scorecards contributes to the measurement of three hypothesis, namely, *HJ*, *H3* and *H9*. The scorecard is subdivided into three sections, namely, demonstration characteristics, demonstration performance and account functionality.

7.3.2.1 Demonstration Characteristics

Demo characteristics are evaluated using measures illustrated in Table (7.4, next page) and test,

> ***H 1: Purchase intentions, mediated by the perceived information search and experience quality, are influenced by the contents and presentation of the demonstration component of an Internet banking service.***

The scorecard contain three groups of measures, content, containing five measures and design and design features of the demo containing a further seven measures. The content and the information on the demo play a crucial part in assisting the information search (Gehrke and Turban, 1999; Bell and Tang, 1998). The measures are designed so that they evaluate the degree to which a consumer may be able to enhance their knowledge on the delivery aspect of the Internet banking service offered by the bank (Schillervaert, *et al.*, 1998).

The emphasis in the group of measures under the design aspect is to evaluate how the demo is presented. The evaluation take a similar approach descried above in the www scorecard. The use of colours, graphics and layout, aspects such as use of frames and tables, likelihood of distortions under a multitude of screen specifications and settings, etc. are evaluated (Bauer and Scharl, 2000; Bell and Tang, 1998; Gehrke and Turban, 1999; Bell and Tang, 1998). A noteworthy additional measure used here, but not used in the www evaluation, is the evaluation of additional multimedia software requirements Haine, (1998).

	Title/Measure	Explanation	Theoretical Support
General Characteristics			
1	Clarity	The demo operation is clear and unambiguous	Gehrke and Turban, 1999; Bell and Tang, 1998
2	Relevancy	Demo is relevant to the context.	Davis, 1989; Doll and Torkzadeh, 1988; Muylle, *et al.*, 1999, Bell and Tang, 1998
3	Comprehensibility	The demo operation is easy to understand and operate.	Doll and Torkzadeh, 1988; Muylle, *et al.*, 1999
4	Comprehensiveness	The demo conveys all information necessary. The need for clarifications is minimised.	Doll and Torkzadeh, 1988; Muylle, *et al.*, 1999
5	Format	The format adopted for the demo is suitable.	Jayawardhena and Foley, 2000
Design			
1	Layout	The demo layout is appropriate.	Doll, et al., 1994; Muylle, *et al.*, 1999; Bell and Tang, 1998;
2	Frames	The demo uses frames. Low reliance is indicated with a high score. Heavy reliance leads to distorted pictures with different Strauss and Frost, 1999; screen settings, downloading can take longer, more prone to errors if the Internet connection is not stable/fast, makes book marking difficult.	Gehrke and Turban, 1999; Sterne, 1999; Sterne and Priore, 2000.

3	Tables	The demo and any explanatory notes are arranged in tables/column format reducing the need for frames. It is also a useful feature to incorporate so that different screen settings can be accommodated.	Gehrke and Turban, 1999; Strauss and Frost, 1999; Sterne, 1999; Sterne and Priore, 2000.
4	Use of colours	Choice of colours is easy on the eye, attractive and is appropriate.	Gehrke and Turban, 1999; Strauss and Frost, 1999; Sterne, 1999;
5	Use of graphics	The demo contains simple and meaningful graphics that are appropriate.	Bell and Tang, 1998; Gehrke and Turban, 1999; Bell and Tang, 1998;
Other Design Features			
1	Animation/multimedia plug in requirements	Animation/plug-in software is required to operate the demo. These lead to slow page loading, hence slows down the operation of the demo.	Gehrke and Turban, 1999; Haine, 1998; Sterne, 1999; Bauer and Scharl, 2000; Janal, 2000
2	Number of screens	The number of Web pages in the demo.	Janal, 2000; Sterne, 1999

Table (7.4): Demo characteristics

7.3.3.2 Demonstration Performance

This scorecard assists in the testing of,

H 9: Purchase intentions, mediated by perceived Internet service quality, are influenced by the performance of the Web site and related components.

In most cases the performance of the demonstration is dependant upon the performance of the Web site. Hence, where applicable, the evaluations made under the demo performance scorecard are read in conjunction with the evaluation of the www performance scorecard. This will be clear in section (8.4.1.1).

	Title/Measure	Explanation	Theoretical Support
Operation			
1	Animation/multi-media plug in requirements	Animation/plug-in software is required to operate the demo and account. These lead to slow page loading, hence slows down the operation of the demo.	Gehrke and Turban, 1999; Bell and Tang, 1998; Bauer and Scharl, 2000; Haine, 1998.
2	Additional software	Additional bespoke software is needed to operate the demo/account. This may consist of automatically downloading Java applets that reside in the computer	Gehrke and Turban, 1999; Bell and Tang, 1998; Bauer and Scharl, 2000; Sterne, 1999; Jana), 2000
3	If so, uninstall facility?	If bespoke software downloads are needed, these can be uninstalled.	Gehrke and Turban, 1999; Bell and Tang, 1998; Bauer and Scharl, 2000; Sterne, 1999; Janal, 2000
4	Off line operation	The information residing in the computer, those related to the account, can be accessed without establishing a connection (going online) to the bank.	Bauer and Scharl, 2000; Janal, 2000
5	Platform freedom	The demo/account can be operated on different platforms	Bauer and Scharl, 2000; Jana, 2000
Demo speed			
1	Downloading time (min)	Time taken for downloading the software, if applicable	Gehrke and Turban, 1999; Bell and Tang, 1998; Bauer and Scharl, 2000;

Table (7.5): Demo operational performance

As shown in Table (7.5), six measures are taken to evaluate the operational performance of the demonstration and by extension the predicted performance of the account. These measures are divided into two groups. Operation of the demonstration and the account encompasses all the requirements that the consumer and their computer have to comply to operate the account. These encompass aspects of the need for additional software (Bauer and Scharl, 2000; Haine, 1998), ways in which software have to be configured, particular ways in which the browsing process should be undertaken (Gehrke and Turban, 1999), conditions under which information is sent over the Internet (Bell and Tang, 1998), the ability to operate in offline mode and platform dependency (Gehrke and Turban, 1999; Bauer and Scharl, 2000). Finally, downloading aspects are evaluated (Gehrke and Turban, 1999; Bell and Tang, 1998; Bauer and Scharl, 2000).

7.3.2.3 *Account Functionality*

The demo together with other information on the Web site conveys the functionality of the Internet banking service account, and tests,

> ***H 3: Purchase intentions, mediated by perceived technical qualities, are influenced by the functionality of banking service products of an Internet banking service.***

This element is perhaps one of the most important aspect for many consumers. Thirty-eight measures are evaluated, divided into eleven groups of measures. The list of measures is not illustrated here for it would merely duplicate a long list of measures that are adequately explained in section (8.3.2). Two types of accounts, current accounts and savings accounts, are evaluated for the purposes of the research. The functionality measures would have to be expanded if a larger selection of the types of accounts were researched.

7.3.3 Communication Methods Mystery-Shopping Evaluation

The amount of information that can be provided on an Internet banking service Web site is finite. Providing too much information may lead to an extremely large Web site, which may put off consumers, because it will make finding the required information an extremely laborious task. There is also the danger that a complicated and overly large site will simply be too complex for most consumers and some consumers will never find what they are looking for at all. The maintenance costs of a large Web site can be comparatively expensive, since many resources may be needed for its upkeep. Therefore most Web sites contain a 'limited' amount of generic information. This means that an instrument or a process must be in place to cater for a consumer's specific query with a tailored response. In almost all cases an appropriate response for such questions require a two-way channel of communication to be established within the WWW. Many Web sites provide an email address, or an online query form on the Web site. Replies to these types of queries are usually sent to the consumer's personal email account. In addition, once the consumer establishes an Internet banking service account, there will be occasions that it will be necessary to communicate with the bank, hence there is a need for an Internet based communication channel.

Therefore, the email/query form mystery-shopping evaluation scorecards contributes to the measurement of two hypotheses, namely that purchase intentions are influenced by, *H9* and *H10*. In order to generate a bespoke response from each of the Internet banking services an email or online query (as appropriate) with the following text was sent to each bank.

I am interested in opening an Internet banking account with your bank. Having browsed through your Web site, I still have a few questions that need clarification.

1. *I am a bit worried about the security. If a third party gains access to my account (perhaps by 'hacking') and carries out an unauthorised transaction, will I be indemnified by the bank?*
2. *In case of the failure of the system, how can I carry out an urgent transaction? Will I incur any additional charges in such an event?*
3. *From time to time I get cheques in US$. Is it possible to deposit these into my account? If so, how do I deposit them? What, if any, commission is charged for this service?*

The response to this query is evaluated using nine measures and is found in Table (7.7, next page). The measures evaluate characteristics of the mode of communication, aspects of speed, responsiveness and by extension the competence of the banking organisation (Avikiran, 1994 and 1999; Parasuraman, 1988 and 1991; Khazeh and Decker, 1992; Johnston, 1997; Levesque and McDougall, 1996).

7.3.3.1 *Characteristics*

The measures on Table (7.6) evaluate performance characteristics of the communication method. These assist in the testing of,

H 9: Purchase intentions, mediated by perceived Internet service quality, are influenced by the performance of the Web site and related components.

General characteristics involve the evaluation of the type of communication method and its security level. It also to ensure that such methods function in an intended manner and consumers find it easy to comprehend such measures (Sterne and Priore, 2000).

	Title/Measure	Explanation	Theoretical Support
General			
1	Availability	Is there a facility available to contact the bank? If not facility is available an email is sent to the postmaster with domain extension (P)	Sterne and Priore, 2000;
2	Type	If available, what type of facility is it? Web based query form/ application (W) or an email link (E)?	Sterne and Priore, 2000;
3	Secure?	When sending/requesting information is the connection secure?	Sterne and Priore, 2000;
4	Bug free	The Web interface/ program (for communication) does the job intended the first time, and correctly. No software 'bugs'	Levesque and McDougall, 1996; Khazeh and Decker, 1992; Johnston, 1997
5	Ambiguity	The Web interface program (for communication) is unambiguous	Sterne and Priore, 2000;
Other Features			
1	News Distribution	The degree to which the bank keep customers effectively up to date with important news through email or message 'pop ups'.	Avikiran, 1994 and 1998; Parasuraman, 1988 and 1991

Table (7.6): Communication methods performance characteristics

7.3.3.2 Performance Characteristics of Communication Methods

The measures on Table (7.7, next page) evaluate performance characteristics of the communication method. These assist in the testing of,

> ***H 10: Purchase intentions, mediated by perceived Internet service quality, are influenced by the responsiveness of communication, speed and accuracy of responses to customers.***

Two of the biggest advantages of an Internet banking service, as far as a consumer is concerned, are the availability of banking services at any time and the near instantaneous speed of execution of instructions. A speedy reply to a consumer query can serve as a surrogate indicator Internet banking service quality (Avikiran, 1994 and 1999; Parasuraman, 1988 and 1991). It is also an indicator of the speed an Internet banking service will put right any mistake that may be made by the bank (Avikiran, 1994 and 1999). Therefore the evaluation of these attributes is important.

	Title/Measure	Explanation	Theoretical Support
Speed and accuracy of response			
1	Email/query sent on (First Round)	Time and date (xxxx - hrs, xx/xx - day/ month)	Sterne and Priore, 2000;
2	Reply Received (First Round)	Time and date (xxxx - hrs, xx/xx - day/ month)	Sterne and Priore, 2000;
3	Prompt response	Promptness of service regarding query - tested through email inquiry reply time gap	Avikiran, 1994 and 1998; Parasuraman, 1988 and 1991; Khazeh and Decker, 1992; Johnston, 1997; Levesque and McDougall, 1996
Responsiveness of Communication			
1	Useful response	The answer to a query is useful, it answers the question adequately.	Avikiran, 1994 and 1998; Khazeh and Decker, 1992; Johnston, 1997; Levesque and McDougall, 1996

2	Understanding	The email reply service demonstrates the ability to Avikiran, 1994 and 1998; understand the specific needs of customers.	Avikiran, 1994 and 1998; Parasuraman, 1988 and 1991; Johnston, 1997; Levesque and McDougall, 1996
3	Sympathy	The level of sympathy inherent in an email response when a customer has problems.	Avikiran, 1994 and 1998; Parasuraman, 1988 and 1991; Johnston, 1997; Levesque and McDougall, 1996
4	Customer interest	The bank's Internet banking operation has a customer's interest at heart (perception measured through experiencing the entire email/communication operation).	Avikiran, 1994 and 1998; Parasuraman, 1988 and 1991; Khazeh and Decker, 1992; Johnston, 1997
Competence			
1	Competence and adequacy of response	The response addresses the question(s) raised in a competent and adequate manner. The language used is adequate and appropriate.	Avikiran, 1994 and 1998; Johnston, 1997; Levesque and McDougall, 1996
2	Error free	The degree of accuracy of a response received by a customer.	Avikiran, 1994 and 1998; Parasuraman, 1988 and 1991; Johnston, 1997; Levesque and McDougall, 1996

Table (7.7): Communication methods performance characteristics

Further, the time lapse between request for information and response is evaluated. Once this process is complete then an analysis of the responses is undertaken to evaluate the responsiveness and competence of the Internet banking service.

Assurance in terrestrial banking refers to the knowledge and courtesy of employees and their ability to inspire trust and confidence. Empathy pertains to the caring, individualised attention that a bank may provide for its customers. Avkiran (1999) defines credibility to signify the maintenance of staff customer trust by rectifying mistakes (Avikiran, 1994 and 1998), keeping customers informed, and communication as the process of fulfilling the banking needs of customers by successfully communicating financial advice and serving timely notices (Khazeh and Decker, 1992; Johnston, 1997; Levesque and McDougall, 1996). These dimensions are quite difficult to offer through the Internet.

Under the broad heading of responsiveness, three interrelated aspects are measured using ten evaluative criteria. The response to an electronic communication is evaluated to determine whether it demonstrates assurance and empathy for consumers (Avikiran, 1994 and 1998; Parasuraman, 1988 and 1991). As stated earlier, empathy pertains to the caring, individualised attention that a bank may provide to its customers. Such responses are also evaluated for competence. Incidentally, the speed of response is itself an indication of the competence of the back office competence in dealing with consumer queries and problems.

7.3.4 Overall Mystery-Shopping Evaluation

Having evaluated the Web site, the demonstration, the email/query procedure, it is now opportune to carry out an evaluation of measures that are not captured by these components of an Internet banking service. Price and branding are aspects that have not been evaluated in the above series of Internet scorecards. Therefore, the overall mystery-shopping evaluation scorecards contributes to the testing of four hypotheses, namely, *H4, H5, H6* and *H7*.

The overall mystery-shopping series of scorecards were subdivided into five sections, namely, current account features, savings account features (Leonard and Spencer, 1991; Jain, *et al.*, 1987), brandings aspects (three scorecards each dealing with separate hypothesis) (Arora, *et al.*, 1985; Laroche and Manning, 1984; Devlin *et al.*, 1995; K.hazeh and Decker, 1992) and overall security perceptions.

7.3.4.1 Current Account Features

Consumers require different bank accounts to perform different functions. Apart from the functionality of the accounts, which have already been evaluated, there are other aspects that are equally important in the decision-making process leading to a purchase. In the literature review, the price of a service was seen as a crucial element in the decision to purchase, and an Internet banking service is no exception. Therefore all price-related aspects of a current account in an Internet banking service are evaluated using twelve measures and are shown in Table (7.8).

7.3.4.2 *Savings Account Features*

Similar to the evaluation of the current accounts, savings account attributes are evaluated using six measures, and are shown in Table (7.9). Taken together, the measures in Table (7.8) and Table (7.9) would be utilised to test,

H 4: Purchase intentions are influenced by the prices (as measured by interest rates, fees, etc.) of an Internet banking service.

	Title/Measure
	Account Attributes
1	Cheque books and payment of cheques
2	Payments into the account
3	Issue debit cards
4	Paying standing orders and Direct Debits
5	Withdrawals from ATMs
6	Preparing and sending out statements
7	Stopping cheques (£)
8	Cash over the counter
	Interest Rates
1	Interest for positive balances (Gross %)
	Overdrafts - Authorised
1	Interest rate (in %)
	Overdrafts - Unauthorised
1	Interest rate (in %)
2	Safety ceiling (in £)

Table (7.8): Current account price sensitive features

	Title/Measure
	Withdrawals
1	From ATMs
2	From branch
	Deposits
1	Through branch

2	Through post office
	Statements
1	Monthly statements
	Interest rate (in %) for £1000
1	Interest rate (in %)

Table (7.9): Savings account price sensitive features

7.3.4.3 *Branding and Related Aspects*

Branding and its related aspects are evaluated using three separate Internet scorecards. The first scorecard deals with visibility.

Visibility

The importance of branding and its influence on consumer purchase intentions were illustrated in the literature review (Arora, *et al.*, 1985; Laroche and Manning, 1984; Devlin *et al.*, 1995; Khazeh and Decker, 1992).

The evaluation of visibility is divided into four groups of measures, namely, (1) online visibility, (2) terrestrial visibility, (3) Web traffic volume and (4) reviews.

In the first group of measures, Internet (online) visibility is evaluated by the search engine capture of the Web sites by the ten most popular Web search engines. The ten most popular search engines in the Internet are selected by referring to an independent body, "searchenginewatch", which monitors the popularity of search engines. Based on these rankings the ten most popular search engines in the Internet during January 2000 were;

1. Yahoo - http://uk.yahoo.com/
2. Altavista - http://www.altavista.com/
3. Excite - http://www.excite.com/
4. Go - http://www.go.com/
5. Lycos - http://www.lycos.co.uk/
6. Snap - http://www.snap.com/
7. Goto - http://www.goto.com/
8. MSN - http://search.msn.com/
9. WebCrawler - http://www.Webcrawler.com/
10. Ask Jeeves - http://www.askjeeves.co.uk/

Within this group, two measures are employed to evaluate search engine capture. Since there are ten search engines these two measures are in effect transformed into two groups of ten measures each. In the first ten measures, a search string is keyed into the search engine and the results are recorded. For this purpose the following search strings are used, 'online banking in UK', 'Internet banking in UK', 'UK Internet banks', each search string is input in three methods; (a) as typed above, (b) with the term AND between words and (c) with '+' between words. If for either of these search strings the bank URL is given as output, it is recorded as a capture. In the next set of ten measures, the name of the Internet banking service is keyed into the search engine. The capture is recorded as above described.

The second measure of visibility is evaluated by simply denoting whether or not the bank has a terrestrial branch network.

The third group of measures of evaluate the reputation. The first measure counts the number of hyperlinks to an Internet bank from other Web sites. Such hyperlinks are also an indicator of how easily a consumer may find their way to the bank Web site from other 'related' Web sites. Related in this context refers to a very wide range of Web sites, including, portals dealing with financial services, other financial institutions, discussion groups and bulletin boards, etc. This count is measured by using the Alexa monitor. As previously stated, this software robot (or agent) is constantly scouring the Internet collecting information, and whenever hyperlinks from a Web site match those of other URL domains already in its database, it records that as a distinct link. The remaining objective measures are designed to evaluate the Web traffic volume or the number of Web users using the site. This is used as a surrogate for the popularity or the reputation of the Web site. The data are recorded from the Alexa monitor.

The fourth group of measures evaluate reviews of the Web site and its contents from independent bodies and discussion groups.

All four groups measures are shown in Table (7.10, next page) and they assist in testing,

> ***H 5: Purchase intentions are influenced by the visibility (as measured by capture by popular search engines and linkages to other Web sites) of a brand name of an Internet banking service.***

	Title/Measure
	Online Visibility
10	**Search engine capture - Search String** (measured for the 10 most popular search engines identified above) - **10 measures**

10	**Search engine capture - Name Search** (measured for the 10 most popular search engines identified above) - **10 measures**
	Terrestrial Visibility
1	Terrestrial branch network
	Web Traffic Volume/Reputation
1	Number of links from other web sites (Alexa count)
2	Traffic (Monitored by Alexa)
3	Alexa Visits
4	Pages (Counted by Alexa)
	Reviews
1	Britannica
2	Alexa Grading
3	Moneybags

Table (7.10): Brand visibility measures

Association and Experience with Brand

The literature review recognise that if a new Internet banking service demonstrates its relationship with a parent bank, which has a strong brand presence, this association was perceived to mitigate risk perceptions of consumers. Similarly, experience with the parent product provides consumers with information about the new product; this might reduce consumer uncertainty and, thus perceived risk associated with the Internet banking service. As the 'new' Internet banking service brand becomes established, consumer risk perceptions may subside. These aspects are evaluated using five measures, and are shown in Table (7.11, next page). While the measures evaluating the association with brand tests,

H 6: Purchase intentions are influenced by the association (or proximity) of an old established umbrella brand.

and, measures evaluating experience with brand tests,

H 7: Purchase intentions are influenced by the strength of a brand, as proxied by the time elapsed since brand launch.

	Title/Measure
	Association of brand
1	Association
2	Stability

3	Network of branches
	Experience with Parent Brand
1	Date of web site activation
2	Launch of transactional Internet banking service

Table (7.11): Association and experience with brand

7.3.4.4 Account Security Features

Security is one of the most important, if not the most important aspect of a bank account that any consumer is concerned about, and Internet banking services are no exceptions. If anything, security concerns on the Internet are even higher, although large-scale security problems with Internet banking services or electronic commerce are yet to emerge. The media gives extensive coverage to even relatively minor breaches of security in Internet banking services or electronic commerce. The evaluative measures are shown in Table (7.12, next page), and they assist in testing of,

H 11: Purchase intentions, mediated by perceived Internet service quality, are influenced by Web site security.

Fourteen measures are taken to evaluate security aspects of the account, and are divided into three groups. Eleven evaluation criteria measure the security measures used in protecting customer accounts. These measures are formulated to evaluate all aspects of security, ranging from software to hardware (Ho, 1997; Muylle, et al., 1999; Bauer and Scharl, 2000). The second group of measures deals with how the bank attempts to instil confidence and build trust with consumers through their Web site, and ultimately how these will work to the consumer's benefit when they open an account (Avikiran, 1994 and 1998; Parasuraman, 1988 and 1991; Khazeh and Decker, 1992). And finally the third group of measures evaluate the remedies in place to protect the consumer, should anything contrary to the intended course of events take place (Johnston, 1997). In this context the regulatory regime under which the individual Internet banking services operate is recorded. This has important ramifications if disputes, bankruptcies, etc. arise in the course of the banking relationship, and perhaps most importantly play a major role in the security perceptions of a consumer.

	Title/Measure	Explanation	Theoretical Support
Security Measures Used			
1	Pages with secure connection	Pages that require secure connection are present. These contribute to the security perception.	Ho, 1997; Muylle, *et al.*, 1999; Bauer and Scharl, 2000; Economist, 2000;
2	Encryption	Level of encryption, measured by bit size. Information found on Web site.	Ho, 1997; Muylle, *et al.*, 1999; Bauer and Scharl, 2000; Economist, 2000;
3	Password	A password is required to gain access to the account.	Bauer and Scharl, 2000; Economist, 2000;
4	Other unique customer identifiers (number)	How many other unique customer identifiers (pieces of information) required operating the account? Added security features employed to secure account service, identified by only the customer.	Sterne 1996 and 1999; Chaffey, *et al.*, 2000
5	Unique transaction numbers	A list of random computer generated numbers, to be used before a transaction, or used after a transaction to confirm it (accessible only to the customer)	Economist, 2000; Sterne 1996 and 1999; Chaffey, *et al.*, 2000
6	General explanation	The explanation of security measures in place is comprehensive.	Sterne, 1996; Jana), 2000

7	Overall security perception	Overall security perception	Avikiran, 1994 and 1998; Parasuraman, 1988 and 1991; Khazeh and Decker, 1992; Johnston, 1997; Sterne 1996
8	Confidence	The Internet banking service, comprising of the Web site, the demo, the functionality, etc. taken as a whole instill confidence.	Avikiran, 1994 and 1998; Parasuraman, 1988 and 1991; Khazeh and Decker, 1992; Sterne, 1996
9	Lockout	Forces the user to start a new session, while in security zone due to time lockouts, password and other security lockouts.	Ho, 1997; Muylle, *et al.*, 1999; Bauer and Scharl, 2000;
10	Secure communication	When sending information, a secure connection is established.	Ho, 1997; Muylle, *et al.*, 1999; Bauer and Scharl, 2000;
11	BS for Security	The security measures in place comply with BS7799 for Information Security Management	
Language			
1	Explicit	Explicit statements on the kind of technology used.	Johnston, 1997; Sterne and Priore, 2000
2	'Small print'	Clear and concise statements on security aspects of the account operation. Minimum amount of disclaimers.	Johnston, 1997; Sterne and Priore, 2000; Sterne, 1999
Regulatory			
1	Country of banking license	The country of jurisdiction, should a dispute arise.	

Table (7.12): Account security measures

While this Internet scorecard contains the primary evaluative criteria, some aspects of security are evaluated in the email evaluation scorecard.

7.4 A Cross Mapping Exercise

Hypotheses are fundamental to the central research question. Therefore it is important to demonstrate that all hypotheses can be robustly tested using a well-founded research methodology. The previous chapter proposed the mystery-shopping methodology and this chapter operationalised the methodology by illustrating the research instrument, i.e. a series of Internet mystery-shopping scorecards.

As one would expect with hypotheses developed to examine an aspect such as purchase intentions, the hypotheses cover a number of academically distinct areas of study but at the same time it has been shown that there are many linkages between these areas of study. In adopting this process, the final tally produced thirteen interrelated Internet scorecards. Altogether these scorecards contain 180 evaluative measures in total. The best method to illustrate these interlinkages and complementarities is in a table that cross references evaluative measures with hypotheses. For ease of re-examination, the relevant sections explaining how key measures were derived are also included in Table (7.13).

Hypothesis	Evaluative measures (Table Number)	Chapter section
H 1: Purchase intentions, mediated by the perceived information search and experience quality, are influenced by the contents and presentation of the demonstration component of an Internet banking service.	Table (7.4)	(7.3.2. I)
H 2: Purchase intentions, mediated by perceived information search quality, are influenced by the contents of an Internet banking service Web site.	Table (7.1)	(7.3.1.1)

H 3: Purchase intentions, mediated by perceived technical qualities, are influenced by the functionality of banking service products of an Internet banking service.	Table (8.5) -the measures are not listed in this chapter for brevity. The measures are self-explanatory.	(7.3.2.3)
H 4: Purchase intentions are influenced by the prices (as measured by interest rates, fees, etc.) of an Internet banking service.	Table (7.8) and Table (7.9)	(7.3.4.1) and (7.3.4.2)
H 5: Purchase intentions are influenced by the visibility (as measured by capture by popular search engines and linkages to other Web sites) of a brand of an Internet banking service.	Table (7.10)	(7.3.4.3) Visibility
H 6: Purchase intentions are influenced by the association (or proximity) of an old established umbrella brand.	Table (7.11)	(7.3.4.3) Association and experience with brand
H 7: Purchase intentions are influenced by the strength of a brand, as proxied by the time elapsed since brand launch.	Table (7.11)	(7.3.4.3) Association and experience with brand
H 8: Purchase intentions, mediated by perceived Internet service quality, are influenced by Web site characteristics.	Table (7.2)	(7.3.1.2)

H 9: Purchase intentions, mediated by perceived Internet service quality, are influenced by the performance of the Web site and related components.	Table (7.3); Table (7.5) and Table (7.6)	(7.3.1.3) Web site speed; (7.3.2.2) Demo performance; (7.3.3.1)
H 10: Purchase intentions, mediated by perceived Internet service quality, are influenced by the responsiveness of communication, speed and accuracy of responses to customers.	Table (7.7)	(7.3.3.2)
H 11: Purchase intentions, mediated by perceived Internet service quality, are influenced by Web site security.	Table (7.12)	(7.3.4.4)

Table (7.13): Cross mapping of hypotheses to chapter sections and Tables containing evaluative measures

7.5 Mystery-Shopping of Internet Banks

Data are collected using the mystery-shopping survey instruments, described above, on all of the Internet bank service providers in the UK as of 1st of January 2000. There were fifteen Internet banks offering a variety of banking products and services. Apart from using the Internet as a distribution channel, some of these banks had up to three distribution channels. The composition is as follows. Three stand-alone Internet banks allowing consumers to bank only over the Internet; two banks which allow consumers to bank using both the Internet and the telephone; ten remaining banks that are traditional banks with extensive branch networks, allowing consumers to bank through a combination of branch, telephone and the Internet. Table (7.14) gives a detailed composition of these banks.

Web site pages from all fifteen banks were downloaded in the late December 1999 for analysis. To reiterate, the purpose of this exercise is to ensure that all Web pages, emails, letters and other communication are recorded for subsequent replication and reinvestigation. A descriptive account of these Internet banking services is found in Appendix (A).

	Name of bank	Distribution Channel(s)		
		Internet	Telephone	Branch
1	BankNet Electronic Banking Service	x		
2	first-e	x		
3	Smile	x		
4	Egg	x	x	
5	First Direct	x	x	
6	Bank of Scotland (BOS)	x	x	x
7	Barclays	x	x	x
8	Citibank	x	x	x
9	Co-operative bank	x	x	x
10	Halifax online	x	x	x
11	Lloyds TSB online	x	x	x
12	Nationwide Online Banking	x	x	x
13	Natwest Bank	x	x	x
14	Norwich and Peterborough Building Netmaster Society (NPBS)	x	x	x
15	Royal Bank of Scotland (RBS)	x	x	x

Table (7.14): The composition of the banks in the survey

7.6 Chapter Summary

This chapter has operationalised the research methodology by developing evaluative measures for the research instrument. The sources of information utilised in the development process of measurement instruments and their suitability validity in the use of academic research has been examined. Throughout the chapter, cross-references are made to hypotheses, so that the research instrument maintains the validity of the research. Now that the data collection instruments have been prepared, and the data collected, the next chapter will carry out an exploratory analysis of the collected data to find the factors influencing consumer purchase intentions in Internet banking services.

Chapter 8.

An Exploratory Analysis

8.1 Introduction

This chapter achieves two objectives. Firstly, by carrying out an exploratory analysis it demonstrates what can be achieved with the limited amount of data available, and also indicates the potential for the future development of the research. As stated in the introductory chapter, the variables are not robust enough to allow rigorous statistical analysis. Additionally, audited demand statistics are not available. These two factors render full hypothesis testing infeasible at this stage, but nevertheless are still valid. Therefore the findings in this exploratory analysis are limited to indicative trends of influences on purchase intentions. Secondly, it offers a glimpse of the factors influencing consumer purchase intentions in Internet banking services from both traditional banks and new entrants, by presenting tentative findings of the mystery shopping research survey. A method of conceptualising this process is by multiple regressions. As a first step in this direction, the left hand side (LHS) variable in this study of the determinants of purchase intentions, i.e. demand, needs to be established. Thereafter, each of the hypothesised variables on the right hand side (RHS) affecting demand is examined.

The chapter is divided into four sections. The chapter commences by examining the demand for Internet banking services in the UK. Section two carries out an exploratory investigation of the influence of the information search and acquisition process on consumer purchase intentions. The third section examines the influence of experience evaluation qualities on consumer purchase intentions. The final section of the chapter comprises of a discussion that draws together the characteristics of Internet banking services that influence consumer purchase decisions. It finds that brand, price, service quality, time elapsed since launch and security of Internet banking service influence consumer purchase intentions.

The chapter concludes with a summary of the chapter.

8.2 Demand for Internet Banking Services

As of January 1, 2000, the industry offered virtually no reliable estimates of customer demand, i.e. the main LHS variable. In order to procure indicative data on the demand, individually addressed and signed letters, along with self addressed business reply envelopes, were sent to all fifteen banks. A sample letter appears in Appendix (B). The letters were posted just before Christmas 1999, so that they would arrive by the time bank offices opened in the New Year. The following banks replied after reminders were sent.

1. Bank of Scotland (BOS)
2. Barclays
3. First Direct
4. first-e
5. Lloyds On-line
6. Royal Bank of Scotland (BOS)
7. smile

Thereafter, a series of emails were sent to the media/public relations' office of the five remaining banks. For those that still did not respond were contacted via the telephone. At the conclusion of this process information as given in Table (8.1) were collated.

Name of Bank	Number of customers as of Jan 1, 2000	Number of customers joining per week	Launch date of transactional Internet banking service
Bank of Scotland (BOS)	350,000	Information withheld	Nov-99
BankNet	Information withheld	Information withheld	Jan-95
Barclays	500,000	7,000	Jun-98
Citibank	Information withheld	Information withheld	Apr-97
Co-operative bank	Information withheld	Information withheld	Mar-98
egg	800,000	Information withheld	Nov-98
First Direct	5,000	4,000	Nov-99

first-e	Information withheld	Information withheld	Sep-99
Halifax Online	65,000	Information withheld	Oct-99
Lloyds TSB On-line	180,000	Information withheld	Apr-99
Nationwide Online Banking	250,000	Information withheld	May-97
NatWest Bank	100,000	11,900	Nov-99
NPBS (Netmas-ter)	Information withheld	92	Jun-98
Royal Bank of Scotland (RBS)	300,000	Information withheld	Jun-97
Smile	25,000	Information withheld	Oct-99

Table (8.1): Total number of customers using Internet banking services and their launch dates

Apart from the number of customers with the service as of the reference date, it is also important to record the number of customers joining the service per unit period, per week in this instance. This is due to two reasons, (1) some Internet banking services have had a longer period of operation and (2) some banks with PC banking services are converting their services on to the Internet to provide Internet banking services. Therefore both these groups are likely to have accumulated a bigger customer base compared to the new Internet bank entrants. The number of customers joining per given duration is important in levelling out the two advantages explained above.

It can be seen from Table (8.1) many banks abstained from releasing the required information on the grounds of commercial sensitivity. These demand statistics are not supplemented with information from the media, since there appears to be significant discrepancies between different media publications. For example, in January 1999, the Guardian newspaper reported that the total number of PC and Internet banking accounts at the end of 1998 is less than 200,000; a report by Datamonitor in December 1998 reported put the figure at 1,200,000; Fletcher in May 1999 reported 400,000 accounts at the end of 1998. A possible explanation to the varying estimates of the number of 'customers' may be due to the definition of a customer. While an academic definition of a customer is provided in section (1.2.1), a bank may decide to count a customer in any of the following methods.

1. The number of active accounts.
2. The number of active accounts, but an individual with multiple accounts is counted as a single customer.
3. The number of Internet banking service accounts opened, irrespective of whether accounts are active or not. This may be the case, if the bank is attempting to generate favourable publicity.

Therefore it can be seen that depending on the definition adopted, a single bank may have three different customer demand statistics. It is also inexplicable that, although some banks refrained from giving information for this study citing commercial sensitivity, similar information has appeared in the national media.

Several of these demand statistics merit an explanation. It was stated that BOS commenced their Internet banking service in November 1999, and yet they state that 350,000 customers were using their service two months later. BOS operated a PC banking service, from 1997 to November 1999, and converted all those accounts into Internet banking accounts. Therefore BOS have a large customer base that appears inconceivable after just two months. A number of other traditional banks have gone through this conversion process, notably Barclays and NatWest. First Direct launched their Internet banking service in November 1999, and within a space of two months they were only able attract 5,000. Therefore, it is difficult to believe that 4,000 customers are joining per week.

Finally, it appears that the number of customers joining per week is more commercially sensitive information than the total number of customers. A plausible explanation of this could be that it is reasonable to assume that all Internet banks are evaluating each other's Web sites very closely. Hence they are able to evaluate any change in Web sites and changes in the functionality of different Internet bank accounts. Therefore if competitors gain access to the number of customers joining per week, they are able to determine which account and Web site features influence purchase intentions and which ones may not influence purchase intentions.

8.2.1 Implications

The non-availability of audited information on demand levels renders full hypothesis testing impossible at this stage. However, the first step in the suggested methodology, the survey of Web site characteristics and responses to queries is reported in the following sections in a first attempt to measure RHS variables in the analysis of customer purchases. This exercise provides an appreciation of the importance of service quality, pricing and perceived value, branding and risk in Internet banking services.

As it can be seen from Table (8.1), only the following ten banks have provided figures for their customer base as of 01/01/2000.

1. BOS
2. Barclays
3. egg
4. First Direct
5. Halifax
6. Lloyds TSB
7. Nationwide
8. NatWest Bank
9. RBS
10. smile

Hence in this first step, only these ten banks will be subjected to simple quantitative and/or graphical analysis, however all fifteen will be subjected to qualitative analysis wherever possible. In the long term, a methodology for analysing the demand for Internet bank services could be enhanced by introducing further elements, since this exploratory analysis suffers from a number of problems which need to be resolved.

Above all, audited customer demand information is required. Even then however, the methodology may still not result in robust conclusions. This is because, an institution needs time to establish itself, and this is especially so with a banking institution. Out of the new entrants, smile and first-e and, from the 'traditional' banks, BOS, NatWest, Halifax and First Direct all started Internet banking services after September 1999. The reference date of this study, January 1, 2000 is too close to service launches to reasonably estimate the determinants of demand. It follows from this that the methodology can be significantly improved ultimately by examining time series data.

In the long term, a methodology for analysing the demand for Internet banking services would involve five stages.

1. Researchers' surveys of Web site characteristics.
2. The formation of representative panels of customers to develop measures and weightings for the measurement of Web site characteristics and supplier responsiveness to queries.
3. Factor analysis to combine clusters of characteristics
4. The collection of audited data on actual banking sales from Web sites.

5. Full time-series regression analysis to identify significant influences on actual purchase decisions.

As explained above, the following two sections [i.e. sections (8.3) and (8.4)] present an exploratory analysis of mystery-shopping exercise. Where possible, the analysis is buttressed with consumer contributions drawn from discussion forums. Therefore it acts a tool to triangulate the findings within the analysis. It is important to appreciate that in section (7.2.1.1) it was explained that forum contributors could either be customers of one or more Internet banking services, or they may be consumers intent on making an Internet banking service purchase. Therefore, where comments are presented it is ensured that they are *not* drawn from customers who have already purchased Internet banking services, so as to increase the relevance of the analysis.

8.3 Information Search and Acquisition

Faced with the decision to purchase Internet banking services, a consumer requires information to inform product choice. It was argued in section (5.3.1) that a consumer may acquire information from the demonstration component, and it was also perceived to add to the experience qualities of the service. Additionally, it was also suggested that a consumer might acquire information from the contents of a Web site. The following sub sections will investigate these aspects.

8.3.1 The Demonstration Component

It was seen that the demonstration component is an important tool by which consumers can enhance the information search and acquisition process and assist in the understanding of the purchase process and account operating procedures by simulation. Therefore a demo would also provide a short cut to experience qualities and these were hypothesised to influence purchase intentions.

H 1: Purchase intentions, mediated by the perceived information search and experience quality, are influenced by the contents and presentation of the demonstration component of an Internet banking service.

This section only deals with the influence of information search qualities on purchase intentions. The influence of experience qualities gained by evaluation the performance aspects of the demonstration component are examined in section (8.4.1.2).

Name of Bank	No demo	Screen snapshots	Web based HTML	Web based Java	Requiring a download
BankNet	x				
Egg	x				
RBS		x			
Barclays			x		
First Direct			x		
Lloyds TSB			x		
NatWest			x		
NPBS			x		
Citibank				x	
first-e				x	
Hailfax				x	
Nationwide				x	
BOS					x
Co-operative					x
Smile					x

Table (8.2): Different types of demonstrations.

As previously stated in section (6.6.1.2), demonstrations are authored in a variety of ways, this is important for they have implications on how they operate. The different types of demonstrations encountered are listed in Table (8.2).

BankNet and egg do not have demonstrations on their Web site. These two banks have therefore forgone an opportunity to enhance the quality of information dissemination to consumers and provide them with experience qualities, and this is exemplified by the following contributor in the discussion forum.

"..........egg doesn't even have a demo, how in the world can I get to know how the damn thing work? ...."

(bjgabu, iii discussion forum)

The most elementary type of demo is to display screen shots of an account, and these can be pasted on several Web pages. Apart from textual explanations of the operational procedures set alongside the screen shots, this type of demo lacks the resourcefulness of demos that are authored. Where the demo is authored in HTML, they can display a true account interface. In addition, it is also possible to afford the opportunity for the consumer to engage in a mock transaction. Java authored demos take the demo experience a bit further, and allows the automation of a number of features. For instance, in the Cooperative bank demo, a mouse movement commences the demo, and the consumer can, if they desire, click on an icon and observe the demo carry out a transaction on its own, with explanatory notes appearing at each step in the process. Finally, if a demo requires a download, the number of features offered is comparable to a Java based demo.

As shown in Table (8.3, next page), demo characteristics are evaluated under two groups of measures, 'general', containing five measures, 'design' of the demo containing five measures and 'other design features' containing a further two measures. smile and Citibank demos are superior to others in their role of informing the consumer, for they contain a significant amount of explanatory notes on the operation of the account. These demos are also useful in guiding the consumer through the series of operations that they would be expected to perform should they become customers of the respective banks.

Figure (8.1, next page) shows the relationship between demo characteristics (RHS variable) and customer demand (LHS variable). The average score of demo characteristics is indicated along the horizontal scale. There appears to be no conclusive relationship. Of interest is that some banks with poorly assessed demos appear to attract a larger proportion number of customers. However, this observation is in contrast to the comments on discussion forums, which shows that consumers do observe demos before making purchases. It is worthwhile to note that bank demos with low scores are traditional banks, and the best demo, smile, belongs to a new entrant. However, egg, a new entrant has succeeded in attracting the largest number of customers in spite of not having a demo. Multiple regressions would reveal the true influence of demos but these must await reliable LHS data.

	Title/Measure	BOS	BankNet	Barclays	Citibank	Co-operative	egg	First Direct	first-e	Halifax	Lloyds TSB	Nationwide	Natwest	NPBS	RBS	smile
	General															
1	Clarity	8.5	5.0	7.0	8.0	8.5	n/a	7.0	7.0	7.0	7.0	7.0	7.5	7.5	6.5	8.5
2	Relevancy	8.0	4.5	7.0	8.5	9.0	n/a	7.5	7.5	7.5	7.0	7.0	7.5	7.0	6.5	8.0
3	Comprehensibility	7.5	5.0	7.5	8.0	9.0	n/a	7.0	6.5	7.5	7.5	7.5	7.0	7.5	7.0	8.5
4	Comprehensiveness	8.0	4.0	6.0	7.5	8.0	n/a	7.0	7.0	8.0	7.0	7.5	7.5	7.0	6.0	8.0
5	Format	7.0	3.5	6.5	7.0	8.5	n/a	7.0	7.5	7.5	7.0	7.0	7.5	7.5	6.5	8.5
	Design															
1	Layout	5.0	4.5	7.5	8.0	8.5	n/a	7.0	7.5	6.5	7.0	7.0	7.5	7.5	7.5	8.5
2	Frames	7.0	6.0	7.0	8.5	9.0	n/a	6.5	7.5	7.0	7.5	6.5	7.0	7.5	7.0	8.0
3	Tables	6.0	4.0	4.0	8.0	9.0	n/a	6.5	7.0	7.5	8.0	5.0	7.5	7.5	6.5	8.5
4	Use of colours	7.0	4.0	7.0	7.5	7.0	n/a	6.5	8.0	7.0	8.0	6.5	7.0	7.0	7.5	7.5
5	Use of graphics	7.0	3.0	7.0	7.0	7.0	n/a	6.5	7.5	7.0	8.0	6.0	7.0	7.0	7.5	8.0
	Average score	**7.1**	**4.4**	**6.7**	**7.8**	**8.4**		**6.9**	**7.3**	**7.3**	**7.4**	**6.7**	**7.3**	**7.3**	**6.9**	**8.2**
	Other Design Features															
1	Animation/multimedia plug in requirements	y	n	n	n	n	n/a	y	n	n	n	n	n	n	n	y
2	Number of screens	9.0	1.0	7.0	11.0	20.0	n/a	18.0	8.0	15.0	11.0	17.0	10.0	7.0	5.0	10.0

	Notes
	egg has no demo
	Co-operative, First-e. Smile - demos run as java applications, hence they cannot be saved for retrieval
	Some sections of the BankNet account needs a bespoke software application to operate, but the demo runs on the browser

Table (8.3): Demo characteristics

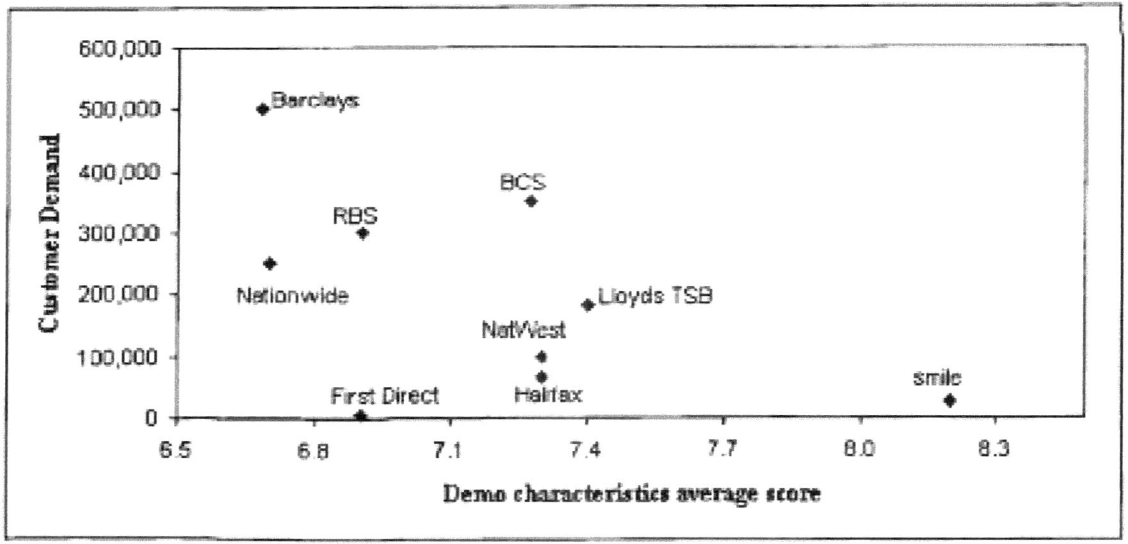

Figure (8.1): Demo characteristics and customer demand

8.3.2 Web Site Contents

As explained above, the contents of a Web site assist in the acquisition of information, and therefore it was hypothesised that,

> ***H 2: Purchase intentions, mediated by perceived information search quality, are influenced by the contents of an Internet banking service Website.***

The first point of contact between a consumer and an Internet banking service is the Internet bank's Web site. The impressions created at this juncture have a significant impact in influencing consumers' purchase intentions. The ability of banks' Web site content to aid information search and acquisition is evaluated using the Internet scorecard shown in Table (8.4).

	Title/Measure	BOS	BankNet	Barclays	Citibank	Co-operative	egg	First Direct	first-e	Halifax	Lloyds TSB	Nationwide	Natwest	NPBS	RBS	smile
	Information/Content															
1	Clear and concise text	7.0	4.5	7.5	7.5	7.0	6.0	7.5	8.0	6.0	6.5	7.5	7.0	7.5	7.5	8.0
2	Relevancy	8.0	3.0	7.5	7.0	6.0	8.0	7.0	7.5	8.0	6.5	7.0	7.0	7.5	7.0	8.5
3	Accuracy	7.5	3.5	7.5	8.5	7.0	7.5	8.0	8.0	7.5	7.5	7.0	7.0	7.0	7.0	8.0
4	Comprehensibility	7.5	5.0	8.0	8.0	7.0	6.0	8.5	8.0	8.0	7.0	7.5	7.5	7.5	7.5	9.0
5	Comprehensiveness	6.0	3.0	8.0	7.0	7.5	5.5	8.0	8.5	8.0	7.0	8.0	7.5	7.0	8.0	9.0
6	Amount of site content	5.5	4.0	8.0	8.0	7.0	6.0	7.5	8.5	8.0	7.0	8.0	8.0	7.0	8.0	9.5
7	FAQs	6.5	4.0	6.0	8.5	6.0	6.5	7.0	7.5	6.5	7.0	7.0	5.0	7.5	9.0	9.0
8	Ease of understanding	5.5	7.0	7.0	6.0	5.0	7.0	7.0	8.0	6.0	6.5	6.0	6.5	7.5	7.5	8.5
9	Disclaimers	5.0	6.0	7.0	5.5	5.5	7.0	7.0	8.5	6.0	6.5	6.0	6.5	7.5	7.5	8.5
10	Format	6.0	3.0	6.0	7.5	5.0	7.5	6.5	8.5	6.0	7.0	6.0	7.0	7.5	7.5	8.5
11	Up to date	7.0	2.0	7.0	5.0	7.0	8.0	7.0	8.0	5.5	5.0	6.0	6.0	6.0	7.5	8.5
	Average scare	6.5	4.1	7.2	7.1	6.4	6.8	7.4	8.1	6.9	6.7	6.9	6.8	7.2	7.6	8.6
12	Amount of downloaded content (no. of pages)	16	11	35	32	24	19	33	32	35	43	26	15	13	19	54

13	Total number of pages on web site - Alexa count	68	515	1	538	119	21	29	14	73	2	108	43	49	81	No data
14	Freshness (Alexa monitor)	5	5	5	5	5	5	5	No data	5	3	5	4	5	5	No data
	Notes Freshness: (5) Less than one week; (4) Less thank 3 months; (3) Less than 6 months; (2) Less thank 12 months; (1) Less than 12 months															

Table (8.4): Web site contents

On most of the Web sites, Web content is easy to understand, comprehensive, useful and is relevant. The exception to this observation is BankNet. This Web site consists of eleven pages, and it does not fully succeed in presenting all the information to the consumer. There are a number of issues not addressed by the BankNet Web site content or are inadequately explained. The NPBS Web site contains thirteen pages. However due to its comprehensive 'Frequently Asked Questions' (FAQ) section it is able to address many concerns and issues that a consumer may have. In general, the FAQ section or similar areas within a Web site attempt to address questions or problematic scenarios that a customer may be faced with. As the name implies, a FAQ section deals only with frequent issues and cannot address every possible query. Hence, where a FAQ section cannot fulfil these requirements, then there is a need for clarification. Such clarifications require individual attention, consuming the bank's resources. Therefore it is in the interest of the banks' to make the FAQ section reasonably comprehensive. In this regard, BankNet and NatWest were adjudged to be comparatively poor while RBS and smile have the most comprehensive FAQ sections.

The number of pages (according to Alexa monitor) includes any page with a hyperlink to the bank Web site, hence the large number of page count in BankNet and Citibank. However where there are frames on a Web site the Alexa count is highly unreliable, for example with the Barclays Web site.

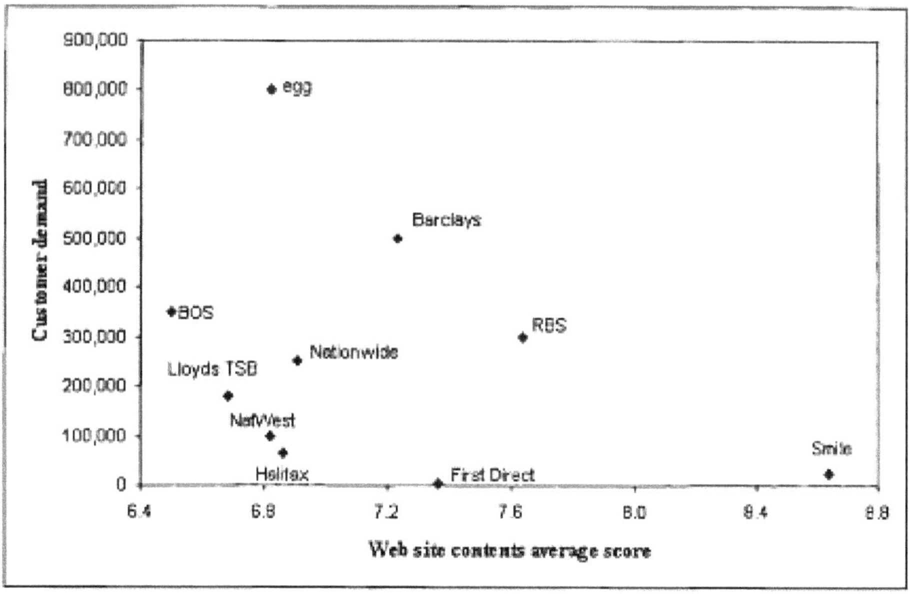

Figure (8.2): Website contents and customer demand

The first-e, RBS and smile Web sites are very effective in their roles as disseminators of information. Additionally, these bank Web sites are also superior to the other Web sites in striking a balance between number of pages and amount of site information. The information found on most of the Web sites, except BankNet and NPBS, appear to have been updated on a regular basis. The last update for pages is April 1999, which in Internet terms is quite old. In general, there was no evidence of conflicting content indicating updating inconsistencies within any of the Web sites. Web content needs to be written in an easy to understand language, comprehensive but there is a trade-off between completeness and brevity (Muylle, *et al.*, 1999; Doll and Torkzadeh, 1988). Although conventional wisdom assumes that too many pages may discourage consumers, a conclusion that can be drawn is that, when assessed in terms of site content and comprehensiveness, there is a positive relationship between length and clarity.

The relationship between (a) a single RHS variable representing Web site contents and (b) customer demand is graphically illustrated in Figure (8.2). The scorecard values (the bi-polar semantic scale values) are converted into an average score, and this is assigned to the horizontal axis. It is inappropriate to carry out a regression of the two variables, due to the incomplete nature of data and is likely to result in spurious conclusions. Upon close visual examination what is evident is that there is no clear relationship between these two variables. It is worthwhile to note that Web site content with low scores are traditional banks, except for one Web site, egg. The best Web site content, smile, belongs to a new entrant. The apparent absence of a clear relationship may be due to other variables that may have a stronger influencing in dictating customer demand.

8.3.3 Functionality

The demo together with other information on the Web site conveys the functionality of the Internet banking service account. Therefore, in section (5.3.1.1) it was hypothesised that,

> ***H 3: Purchase intentions, mediated by perceived technical qualities, are influenced by the functionality of banking service products of an Internet banking service.***

The functionality Internet scorecard contained 38 measures, and the survey results are found in Table (8.5, next page).

Functionality features of an account are one of the most useful aspects of an Internet banking service for a vast majority of consumers. Two types of accounts, current accounts and savings accounts, are evaluated for their functionality. The number of functionality measures on Table (8.5) would increase if more types of accounts, such as mortgage, loan, etc. were included in the research. It can be seen that most banks are already, if not are in the process of offering all major types of account functionality through their Internet banking service. The majority of banks actively encourage consumers to begin the process of opening an account over the Internet. This process saves a great deal of banks' resources and is an ideal way of increasing customer base at minimal cost. The customer perceives that they are in control. Both parties appear to gain from the process. However, consumers are not able to completely conclude the account opening process online. This is because statutory regulations require the banks to physically verify crucial information like the identity of individual(s) and proof of address.

			BOS	BankNet	Barclays	Citibank	Co-operative	egg	First Direct	first-e	Halifax	Lloyds TSB	Nationwide	Natwest	NPBS	RBS	smile
1	**Type of Accounts Covered**	Current	y	y	y	y	y	y	y	y	y	y	y	y	y	y	y
2		Savings	y	y	y	y	y	y	y	y	y	y	y	y	y	y	y
3	**Open Accounts Online**		n	n	y	y	y	y	y	y	y		y	y	y	n	y
4	**State-ments**	View Statements/Account	y	y	y	y	y	y	y	y	y	y	y	y	y	y	y
5		Account History (number of days)	90	30	42	90	180	60	120	Full		60	90		180	180	180
6		Account History (number of transactions)				90					60			40			
7	**Bill Payments**	Current	y	n	y	y	y	n	y	n	y	y	y	y	y	y	y
8		In advance (days)	30		720	120	180		180			360	30	40	720	90	No limit

9		Pay only organisations in "approved list										y		y	y	y
10	**Standing Orders**	View	y	n	y	y	y	n	y	n	y	y	y	y	y	y	y
11		Amend	y	n	n	y	y	n	y	n	y	y	y	n	y	y	y
12		Cancel	y	n	n	y	y	n	y	n	y	y	y	n	y	y	y
13		Set up	n	n	n	n	y	n	n	n	y	n	y	n	n	n	y
14	**Direct Debits**	View	y	n	y	y	y	n	y	n	y	y	y	y	y	y	y
15		Amend	n	n	n	n	n	n	y	n	y	n	n	n	n	y	y
16		Cancel	n	n	n	y	y	n	y	n	y	n	n	n	y	y	y
17	**Transfers**	Transfer funds - within bank	y	n	y	y	y	y	y	y	y	y	y	y	y	y	y
18		Transfer funds - among domestic banks	y	n	y	y	y	y	y	y	y	y	n	y	y	y	y
19		Transfer funds - at a given date	y	n	y	n	n	n	y	y	y	y	n	y	y	y	y
20	**Miscella-neous**	Messages (from bank)	y	y	n	n	n	n	y	y	y	y	y	y	y	y	y

21		Write electronic cheques	n	y	n	n	n	n	n	n	n	n	n	n	n	n	n
22		Accounts amendment	n	n	n	n	y	n	n	y	n	n	y	n	n	n	n
23		Order cheque books	y	y	n	n	y	n	n	y	y	n	y	n	n	n	y
24		One off payments	n	n	y	y	y	y	y	y	y	y	y	y	y	y	y
25		Order/Print statements	n	y	y	y	y	y	y	y	y	y	y	y	y	y	y
26		Request information	n	n	n	y	y	y	y	n	y	y	y	y	y	y	y
27		Calculate Interest accruals	n	n	n	n	n	y	y	n	y	n	n	y	n	n	n
28		Multi currency functions	n	n	n	y	n	n	n	n	n	n	n	y	y	y	n
29		Send Messages	n	n	n	n	y	y	y	y	y	y	y	y	y	n	y

30	**Reconciliation**	Reconciliation/Integration by downloading	n	n	y	y	n	n	y	n	n	y	y	y	y	n	n
31	**Cost**	Cost of operation	Free	Free	Free	Free	Free	Free	Free	Free	Free	Free	Free	Free	Free	Free	Free
32		Cheaper Banking	n	n	y	n	n	n	n	n	n	n	n	n	n	n	n
33		Better/favourable rates	n	n	n	n	n	y	n	n	y	n	y	n	n	n	n
34	**Interactive features**	Budget planner	n	n	n	n	n	n	n	n	n	n	n	y	n	y	n
35		Branch and cash machine locator	n	n	y	n	n	n	n	n	n	y	y	n	n	n	n
36		Time limitations/timeouts	n	n	n	n	y	y	n	y	n	n	n	y	n	n	n
37		Referrals	n	n	n	y	n	n	n	y	n	n	n	n	n	n	n

38	Savings/ Loan/Mortgage calculators	n	n	n	n	n	y	y	n	y	n	n	y	n	y	n
Raw Count		13	7	16	18	20	13	22	14	24	18	21	20	21	21	22

Table (8.5): Functionality of accounts

The ability to view account statements online does not constitute a transaction, but is nevertheless is a very important aspect of an account. All banks afford the opportunity for customers to view the latest statement online. The length of account history that can be viewed online varies from 30 days with BankNet to the full account history (from the day the account was opened) with first-e. An important observation is that although a customer should be able to observe the state of their account as of that moment ('real time'), none of the banks offer this facility. Customers are only able to view the state of their account as of the close of business on the previous working day. This is in part because although Internet banking is a 24-hour banking system with no 'holidays', the accounting system still operates on a traditional weekday basis. There is another reason why these transactions are not processed instantaneously. A bank maintains different processing programs to deal with different types of transactions, and may separate them by geography as well (if it is a traditional bank with branches spread over the UK), so there could be dozens of separate programs running independently of each other. Further, data about transactions carried out are gathered under a 'batch processing' system, in which the information is stored and then sent to the mainframe in batches. The Internet operates in real time, here and now, offering the enormous benefit of continuous monitoring of balances in a variety of accounts. Up to now not a single bank offers the true capabilities of the Internet to customers. It is also important to recognise that because Internet banking services cannot control the speed at which information is sent to the server, they can be susceptible to overloading.

With the exception of BankNet, all those banks offering current account facilities offer bill payments, standing orders and direct debit facilities. Transfers can be broken down into two types, transfers within 'own' bank accounts, and domestic money transfers between banks. Transfer of money within own bank accounts, i.e. where a customer holds several accounts with the same bank, is an important aspect of Internet banking services that can yield practical benefits that are very time dependant. Monies held in current account earn less interest (or no interest at all) than monies held in savings accounts. Hence timely transfer of money within the accounts can yield significant interest payments from the bank to the customer depending on the amounts. Similarly, a customer can exercise more control over their money by carrying domestic money transfers between different bank accounts held with different banks. These accounts may be their own accounts or accounts belonging to third parties. BankNet has no transfer facilities at all, and Nationwide does not allow domestic money transfers. Some banks only allow the transfer process to commence from the moment the instructions are given by the customer, while other banks empower the customer to specify the date on which they want the transfer to take place. Research has found that an average household receives between 10-15 bills a month (Ostrow, 1998), and timely payment of these presents significant savings. Missed payments, even by a few days, carry penalties. To carry out a bill payment, a customer needs to be in possession of the recipient's account number, the sort code and any other reference numbers that may be applicable. As the analysis reveals, four banks, namely Nationwide, NPBS, RBS and smile only allow customers to make payments to an approved list of organizations recognized by the bank. Some customers may be disadvantaged by this limitation, if the recipient is not among those organisations in the bank's list.

Banks also offer a number of miscellaneous facilities. A messaging system is an increasingly used feature. Essentially this system consists of a dedicated Web based mailbox within the customer's account. The customer can use this to send queries, instructions, suggestion, etc. to the bank and the replies will be sent to this mailbox. Further if the bank wants to initiate some bespoke communication or distribute some sort of general news item it may send such messages to the mailbox. The advantage here is that the line of communication is secure, unlike ordinary email received through a 'pop' (post office protocol) type of server.

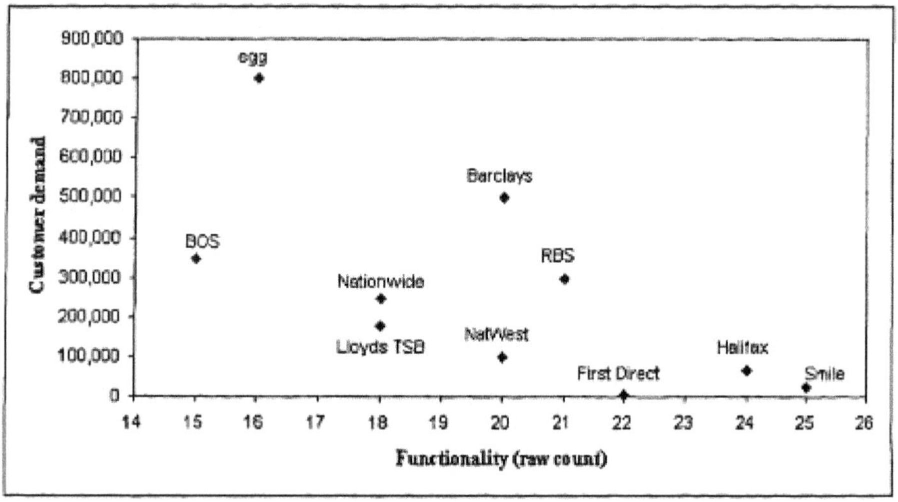

Figure (8.3): Functionality and customer demand

Facilities to make amendments to bank account details, such as the ability to change a customer's mailing address, may appear to be subtle aspects but nevertheless ultimately contribute to the quality of Internet banking service experience. As said earlier, it streamlines the bank administration system, and by delegating these housekeeping tasks to the customer the bank can save resources. Perhaps more importantly the customer perceives that they are in control. Additionally, requisitions for chequebooks, requesting stationary and other information augment the customer empowerment process. On the other hand Web tools such as interest accrual calculations, budget planners, branch and cash machine locators, are very effective practical tools that are of immense use to customers, therefore it is surprising to note that not all Internet banking services offer them.

Consumers increasingly use financial software like Quicken and MS Money for their financial management. A number of banks have ensured that their systems are compatible with financial management software packages, hence the importance of reconciliation facilities.

The relationship between functionality and customer demand is tentatively illustrated in Figure (8.3). The horizontal axis indicates the number of functions available. This raw count is arrived by assigning I and 0 to 'y' and 'n' respectively. Admittedly, this is a crude approximation, since some consumers may perceive that some functions are more important than others. There appears to be no clear relationship. However, it can be argued that an approximate inverse relationship is evident. However, this conclusion does not find resonance with a large number of forum contributors stress the importance of account functionality.

> *"........I'd very much like to start using the internet to run my current account. Not just statements, but being able to do transfers and pay bills ...."*
>
> (Stuham, Motley Fool Discussion Forum)

Therefore, it is difficult arrive at a conclusion with conflicting evidence. True impact must await multiple regression.

A notable feature that is beginning to emerge is that new Internet banking entrants appear to be 'trying harder' to meet customer expectations. This aspect will be explored further in the discussion towards the end of this chapter.

8.3.4 Pricing and Perceived Value in Internet Banking

It has been established that price may have an enormous influence in the shift on consumers from conventional banking services to Internet banking services. Further, it was also argued that price also dictates the shift towards a particular Internet banking service over another Internet banking service. Hence it was hypothesised that,

H 4: Purchase intentions are influenced by the prices (as measured by interest rates, fees, etc.) of an Internet banking service.

However, aspects of price in Internet banking service accounts depend upon the type of account. Accordingly, influence of price on consumer purchase intentions is evaluated under two separate sections, i.e. current and savings accounts. The analysis will begin by scrutinising current accounts, examining prices charged and perceived charges.

8.3.4.1 Current Accounts

This evaluation includes twelve measures, and is shown in Table (8.6). The first eight measures evaluate current account attributes that are not prices that can be expressed in numbers, but nevertheless are perceived as prices (charges) by consumers. The remaining four measures evaluate prices in the form of interest rates on positive balances and prices (charges) relating to overdrafts.

	Title/Measure	BOS	BankNet	Barclays	Citibank	Co-operative	egg	First Direct	first-e	Halifax	Lloyds TSB	Nationwide	Natwest	NPBS	RBS	smile
	Account Attributes															
1	Cheque books and payment of cheques	Free	Free	Free	Free	Free		Free		Free	Free	Free	Free	Free	Free	Free
2	Payments into the account	Free	Free	Free	Free	Free		Free		Free	Free	Free	Free	Free	Free	Free
3	Issue debit cards	Free		Free	Free	Free		Free		Free	Free	Free	Free	Free	Free	Free
4	Paying standing orders and Direct Debits	Free	Free	Free	Free	Free		Free		Free	Free	Free	Free	Free	Free	Free
5	Withdrawals from ATMs	Free		Free	Free	Free		Free		Free	Free	Free	Free	Free	Free	Free
6	Preparing and sending out statements	Free		Free	Free	Free		Free		Free	Free	Free	Free	Free	Free	
7	Stopping cheques (£)			Free						7.5		Free				10.0
8	Cash over the counter	Free		Free	Free	Free		Free		Free	Free	Free	Free	Free	Free	
	Interest Rates															
1	Interest for positive balances (Gross %)	0.1	0.2	2.5	2.5	2.5		0.1		0.3	0.1	1.5	0.1	0.3	0.1	4.8

	Overdrafts - Authorised															
2	Interest rate (in %)	18.0	16.9	18.8	19.0	18.0		16.9		14.3	19.0	11.9	18.4	17.0	19.5	9.9
	Overdrafts - Unauthorised															
3	Interest rate (in %)	29.8	33.5	29.8	29.8	32.9		29.5		29.5	29.8	19.9	29.5	29.9	29.8	22.0
4	Safety ceiling (in £)	100		20		250		250								

Notes

Positive balance interest comparison uses each bank's most similar current account products as at 1 Jan 2000

Barclays - Data is for Prime account

Citibank - Minimum deposit to open an account is £2000

egg - No current account facilities are available

first-a - No current account facilities are available as at Jan 2000. It was introduced in April 2000

first-a - An encrypted statement is sent to the email address

Halifax - Minimum deposit to open an account is £50

Halifax - When a customer takes money out of the Current Account at a cash machine which is not a Halifax or HSBC cash machine, the bank will deduct a charge of £1 out of the account immediately.

Lloyds TSB - Data is for Classic account

NPBS - For balances between £1 to £499

smile - Printed copy of statement - £2.50 - if statement from the last 12 months

smile - Printed copy of statement - £10 - if statement from 1 year ago or over

smile - Withdrawals from all "Link" ATM machines are free

Table (8.6): Current account price features

All fifteen banks, except egg and first-e, offer current accounts. The facilities offered by the banks have already been discussed. Apart from what is already discussed, all banks with current accounts offer chequebooks and ability to pay into the account free of charge. For many consumers the use of a debit card and the facility of withdrawing money from an ATM is synonymous with a current account, all banks except BankNet, offer these facilities. Therefore, this means that a customer with a current account BankNet cannot get cash from their account, which appears to be a rather exceptional restriction. Except for the new Internet bank entrants, i.e., BankNet and smile, all other banks send regular statements to their customers by post. All traditional banks afford the customers the opportunity of withdrawing money from a terrestrial bank branch counter. Customers banking with a new entrant or a bank with a very small branch network, such as Citibank and NPBS, are at a disadvantage, because the amounts of money that can be withdrawn from an ATM is restricted, and cannot withdraw large sums of money, say £1000, in the form of cash. smile is a part of Cooperative bank, First Direct is a subsidiary of HSBC, and hence their customers can also withdraw money from the respective parents. Barclays and Nationwide go to the extent of offering free stop-payment of cheques.

There is a clear demarcation in the interest rates offered between smile, a new entrant, and the other banks. For positive balances, smile offers an interest rate that is almost twice more than the next best rate. Similarly, for authorised overdrafts it is again the bank with the least interest rate. Therefore, as expected, the new entrants offer the most attractive interest rates for all types of accounts, the exception being BankNet.

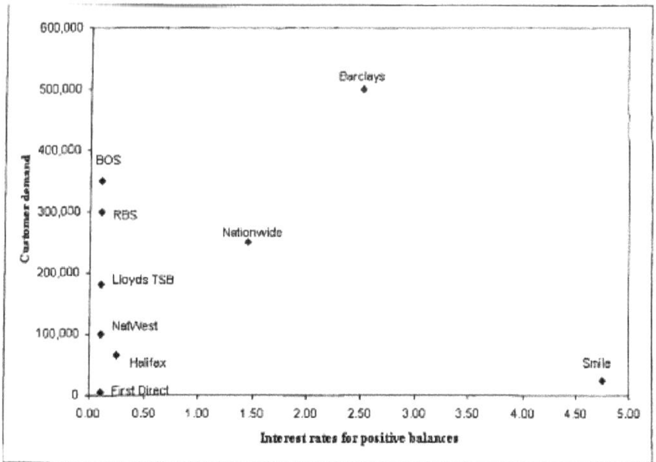

Figure (8.4): Current account interest rates for positive balances and customer demand

The relationship between interest rates for current accounts and customer demand is illustrated in Figure (8.4) and Figure (8.5). Figure (8.4) illustrates the relationship between customer demand and interest paid on current account positive balances. On the face of it there do no appear to be any relationship between the two variables. Of interest is with the exception of Barclays, and to some extent Nationwide. All traditional banks offer extremely small interest (less than 0.3%) on positive balances. Similarly, Figure (8.5), which depicts the relationship between customer demand and overdraft interest charged, does not show a conclusive relationship between the variables. Interestingly, some of the banks that offer the lowest interest on positive balances charge some of the highest interest rates on overdrafts. The observed relationships between customer demand and interest rates cannot be explained in terms of price, since it is logical to expect customer demand to (1) increase with lower overdraft interest rates (low prices), and (2) increase with higher interest rates (low prices) paid on current account positive balances. An examination of forum discussions show that consumers are significantly influenced by account attributes that are less costly (i.e. the low charges) and high interest rates on current account positive balances, for example,

> *"........Overall my I am very impressed at the huge interest rates on CURRENT account at Smile around 4% compared with around .1%for Barclays .... "*
>
> *(Gabriel white, iii)*
>
> *"Like many others I am lured by the promise of low charges.."*
>
> *(24bb65, Motley Fool)*

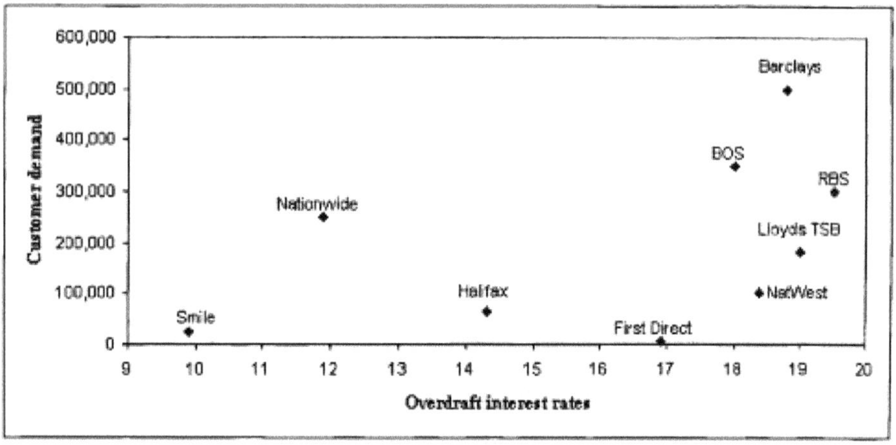

Figure (8.5): Current account overdraft interest rates (authorised) and customer demand

8.3.4.2 *Savings Accounts*

Unlike the current account analysis, savings accounts offer very limited price sensitive account attributes, therefore only interest rates offered is considered for evaluation. All fifteen banks offer savings accounts. All banks except BankNet and first-e, allow cash withdrawals from an ATM, however both Halifax and egg apply a penalty of three quarters of a percentage point in interest rates for this facility. Deposits can be made through the branches and in some instances through the post office.

When Figure (8.6,) is examined, a similar picture to that of current account interest rates begins to appear with regard to savings accounts rates. For instance, all new entrants (except BankNet) offer interest rates above the Bank of England base rate for savings accounts. However, the policy of paying interest rates well above the base rate may not be sustainable in the long run. egg and first-e have guaranteed periods, until such time they will continue to pay interest rates above the base rate. As with virtually all Internet related businesses during 1999, growth is given precedence over profits. Accordingly these stand-alone Internet banks are concentrating on attracting the largest number of customers in the shortest possible time. The rationale being that they can begin to reap the benefits of economies of scale in future, and possibly commence cross-selling of other financial products as described in the second chapter. With customer acquisition costs approaching £150 per customer (Taylor, 1998), it is not surprising to observe that banks such as first-e, First Direct, Citibank, offer monetary incentives to attract new customers. On the other hand, traditional banks with branch networks may waive administrative and documentation charges that would normally be levied on a terrestrial branch consumer.

The combined effects of these incentives, which amount to price reductions and low charges, may have a strong effect on demand. Such demand coupled with publicity from overstating of the number of customers joining Internet banking services could lead on to a bandwagon effect (Moschandreas, 1994).

At the time of writing, none of the banks levy any monthly or other charges to operate an Internet banking service. It is of interest to note that four banks did levy a monthly fee for Internet banking facilities during much of 1999, but it appears that competition from new entrants as described in the second chapter has forced them to abandon monthly charges towards the latter half of 1999. This is a very direct and visible benefit of new entrants.

The association between interest rates paid on savings account balances and customer demand is demonstrated in Figure (8.6, next page).

Eyeballing the Figure it can be seen that there is no relationship and the variables hardly move together.

An important point to note here is that the customer demand has not been differentiated between current accounts and savings accounts. Therefore, apart from the inadequacies associated with demand statistics described and reiterated many times over in this thesis, this is an additional shortcoming. However in the long term, the demand for current accounts and savings accounts must be estimates separately.

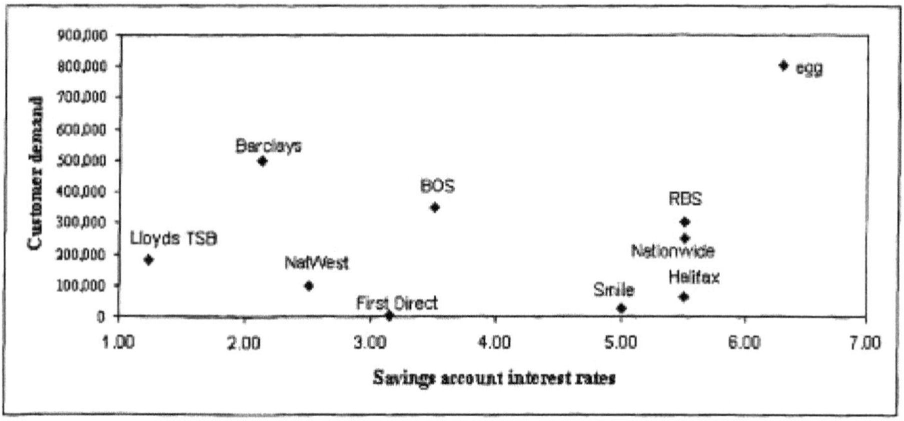

Figure (8.6): Savings account interest rates and customer demand

It is important to examine the customer demand for egg, a new entrant, providing only savings accounts. At time of launch of egg in October 1998, they offered the most attractive interest rate for savings accounts in the UK, with very limited functionality. At that time it was an unknown brand. In spite of these limitations, in the space of fifteen months they have acquired the biggest customer base. Additionally judging by reports appearing in the media, smile and first-e, two net entrants offering comparatively high interest rates, had acquired almost 200,000 customers by June 2000. Further the importance of low prices (i.e. high interest rates) appears to influence discussion forum contributors.

> *"......the only reason why I am even considering banking with first-e is their high interest rates ...."*
>
> (bob78, iii)

On balance, may be concluded that this analysis has shown that, price, has a strong influence on purchase intentions. This may be a dominant variable, however, such a strong conclusion cannot be proven until multiple regressions are carried out.

8.3.5 Brands and Risk in Internet Banking

The influence of brand on purchase intentions was examined in great detail in the previous chapters. It was argued that this influence of brand can be evaluated at three levels, i.e. by assessing the online visibility of the brand, its association with an existing brand (umbrella brand) and time elapsed since launch of brand. Therefore the following sub sections will evaluate each of these aspects of brand.

8.3.5.1 Visibility of Brand

It was hypothesised that,

> ***H 5: Purchase intentions are influenced by the visibility (as measured by capture by popular search engines and linkages to other Web sites) of a brand of an Internet banking service.***

Visibility of brand, is measured on two fronts, Internet brand visibility and terrestrial brand visibility.

While several methods exist to evaluate Internet visibility, the need to select a measure that best reflect the most likely scenario of a consumer browsing the Web in search of an Internet banking service was highlighted in section (7.3.4.3). One of the most prevalent methods by which a consumer would search for an Internet banking service, is to search the Internet for Internet banking services in the UK through a popular search engine. As stated in section (7.3.4.3), ten search engines are selected for this evaluation and the reasons behind the selection of these particular search engines are also presented in that chapter.

The results of the first step, which is indicated by the first ten measures on Table (8.7, next page), where a search string is keyed into the search engine. This process produced results that are quite surprising. The co-operative bank is captured by all but three of the search engines, and at the other extreme BOS, First Direct and RBS are being captured by only one search engine. It is debatable whether this outcome is due to the poor design features of the individual search engine or a function of the Web pages of the particular bank. As stated earlier, most search engines look for Meta tags on Web pages, and omissions of appropriate Meta tags in Web authorship can translate to non-capture by search engines. Further, Web sites may require to be registered with some search engine operating organisations to ensure that they are included in search engine databases. The second step in this search process is to input the name of the Internet banking service into the search engine, i.e. a name search exercise. This process, as shown by the next set of ten measures, ensured that, except for first-e and Bank-Net, all search engines captured all other banks.

	Title/ Measure	BOS	BankNet	Barclays	Citibank	Co-operative	egg	First Direct	first-e	Halifax	Lloyds TSB	Nationwide	Natwest	NPBS	RBS	smile
	Search engine capture·Search String															
1	Yahoo	y	y	y	y	y	y	y	y	y	y	y	y	n	n	y
2	AltaVista	n	n	n	y	y	n	n	n	n	y	y	n	n	n	y
3	Excite	n	n	n	y	n	n	n	n	n	n	n	y	y	n	n
4	Go (Infos-eek)	n	n	n	y	y	n	n	n	n	n	n	n	n	n	y
5	Lycos	n	n	y	n	y	y	n	y	n	n	n	n	y	n	n
6	Snap	n	n	y	n	n	n	n	n	n	n	n	n	n	y	n
7	GoTo	n	n	y	n	n	n	n	n	n	n	n	n	n	n	n
8	MSN	n	n	n	y	y	n	n	n	n	n	n	n	n	n	n
9	Web-Crawler	n	y	n	n	y	n	n	n	y	y	n	n	n	n	y
10	Ask Jeeves.	n	n	y	n	y	n	n	n	n	n	n	n	n	n	n
	Search engine capture·Name Search															
1	Yahoo	y	y	y	y	y	y	y	y	y	y	y	y	y	y	y
2	AltaVista	y	y	y	y	y	y	y	y	y	y	y	y	y	y	y
3	Excite	y	y	y	y	y	y	y	y	y	y	y	y	y	y	y

4	Go (Infos-eek)	y	y	y	y	y	y	y	y	y	y	y	y	y	y	y
5	Lycos	y	y	y	y	y	y	y	y	y	y	y	y	y	y	y
6	Snap	y	y	y	y	y	y	y	y	y	y	y	y	y	y	y
7	GoTo	y	n	y	y	y	y	y	n	y	y	y	y	y	y	y
8	MSN	y	y	y	y	y	y	y	y	y	y	y	y	y	y	y
9	Web Crawler	y	y	y	y	y	y	y	y	y	y	y	y	y	y	y
10	Ask Jeeves.	y	n	y	y	y	y	y	y	y	y	y	y	y	y	y
	Web Traffic Volume/Reputation															
1	Number of links from other Web sites (Alexa count)	405	4,305	3	6,270	372	128	60	No data	532	142	555	280	41	630	No data
2	Traffic (Moni-tored by Alexa)	2	1	3	3	2	3	2	2	1	2	2	2	1	2	2
3	Alexa Visits	1,187	503	11,178	37,240	3,704	27,353	2,524	7,109	No data	1,936	6,940	5,165	389	2,256	6,908

4	Pages (counted by Alexa)	68	515	1	538	119	21	29	No data	73	2	108	43	49	81	No data
	Reviews															
1	Britannica	2	No data	2	2		No data	No data	No data	1	No data	No data	No data	No data	No data	No data
2	Alexa Grading	5	No data	1	1	2	5	No data	No data	No data	No data	3	1	No data	No data	5
3	Money-bags	4	No data	6	7	7	7	7	7		6	8	6	No data	7	9

Notes

Traffic data - (5) Top 100; (4) Top 1,000; (3) Top 10,000; (2) Top 100,000; (1) <Top 100,000

Traffic data is based on "Visits" over the last 6 months, which is a count of the total number of page visits by Alexa Users to this site.

Alexa measures the largest, most geographically and demographically diverse sample of overall Web usage available as of January 2000.

Reviews

Britannica and Alexa: (5) Best of the Web: (4) Highly recommended: (3) Useful; (2) Worth Mentioning: (1) Waste of Time

Moneybags: On a scale of (1) Worst to (10) Best in financial services Web sites

Table (8.7): Brands and on-line visibility

Another measure of visibility is the number of hyperlinks to an Internet bank from other Web sites. Such hyperlinks are also an indicator of how easily a consumer may find their way to the bank Web site from other related Web sites. 'Related' in this context refers to a very wide range of Web sites, including, portals dealing with financial services, other financial institutions, discussion groups and bulletin boards, etc. This count is measured by using the Alexa monitor. As previously stated, this software robot is constantly scouring the Internet collecting information, and whenever hyperlinks from a Web site match those of other URL domains already in its database, it records that as a distinct link. The hyperlink count ranged from 4305, for the most linked, BankNet, and to less than 100 for the least well linked. smile and first-e, did not contain any links, this is probably due to the fact that at the time of the survey, these Web sites had only been in operation for a relatively short period of time compared to the other Web sites. BankNet Web site is under the generic domain name 'mkn.net.', which is part of a relatively large portal dealing with various aspects of shopping and financial services, and this has contributed to the comparatively large number of links from other sites.

A further measure of popularity is the amount of Web site traffic received by a site. Alexa monitor is used to collect data. Barclays, Citibank and Cooperative fall into the world's top 10,000 most visited Web sites; nine others are categorised into the next segment, i.e. top 100,000 sites, the remaining banks, BankNet, Halifax and NPBS, fall outside these categorisations. A fourth measure is reviews of Internet banking service Web sites by independent third parties. Reviews from 'Britannica', 'Alexa' and 'Moneybags' are collected, and these show some variation.

The results on Table (8.7) gives very little evidence to the hypothesis that there is a positive relationship between online visibility and customer demand.

8.3.5.2 Umbrella Brands

As explained in section (5.3.1.3), umbrella branding may influence purchase intentions, hence it was hypothesised that,

H 6: Purchase intentions are influenced by the association (or proximity) of an old established umbrella brand.

In terms of terrestrial visibility, stand-alone Internet only banks may be disadvantaged compared to traditional banks with well-established brands that offer Internet banking services. This is because they lack the everyday high street and other strategic locational visibility through name boards and branch buildings that the latter may enjoy. A link with an established bank or financial institution, more specifically a strong brand, may inspire consumer confidence and trust, thereby associating such banks with stability. The following contributors in discussion forums exemplify these arguments.

> *"..........They are a new bank and I don't recognise the member of the consortium than own them ...."*
>
> (Guavatree, Motley Fool)

> *"....My own preference is for an on-line bank that also has the back -up of access by phone. Nationwide has that and High street presence, stable brand. It may not be the greatest web site, but these other factors matter."*
>
> (Edward William East, iii)

Some new entrant Internet banks that associated with an established brand make the link clear. This was hypothesised to translate to lower risks, and by extension would positively influence purchase intentions. All Internet banking services, except BankNet and first-e, have affiliations of one form or another with a reputed UK financial institution, and therefore they may appear to be stable and trustworthy in the eyes of consumers.

On the other hand, high street visibility may become a burden to traditional banks with Internet banking services from consumers who have had an unsatisfactory conventional banking relationship in the past, and in these instances the association may act as negative influence on purchase intentions.

New Internet bank entrants and their 'new' brands are currently getting an increased proportion of media coverage. Almost all the mainstream national and regional newspapers carry an Internet section either on a daily basis or weekly basis at the minimum. These sections inevitably carry articles describing one or many Internet banking services on a regular basis. Since the emphasis on these newspapers sections are to inform the public, they tend to pay more attention to the new entrants as opposed to the traditional banks offering Internet banking services. Such publicity greatly enhances these Internet banks' visibility and may create bandwagon effects. However the impact of this type of publicity is almost impossible to measure objectively. Additionally, in order to increase brand awareness and increase visibility on the Internet, Internet banks are spending heavily on advertising on both Internet and traditional forms of media. In aggregation the excitement of a new brand may influence consumers in their purchase intentions.

	Title/Measure	BOS	BankNet	Barclays	Citibank	Co-operative	egg	First Direct	first-e	Halifax	Lloyds TSB	Nationwide	Natwest	NPBS	RBS	smile
	Parent Brand Association and Confidence															
1	Association	y	n	y	y	y	y	y	n	y	y	y	y	y	y	y
2	Stability	y	n	y	y	y	y	y	n	y	y	y	y	y	y	y
3	Network of branches	y	n	y	y	y	n	y	n	y	y	y	y	y	y	y

Table (8.8): Umbrella branding

The association with a parent brand is shown in Table (8.8) On the basis of a stable brand, all traditional banks with Internet banking services should possess a large customer base, for these banks may be associated with the least amount of risk. Indeed, with the exception of Halifax, the customer demand for each and every one of these banks is at least 100,000 customers. Customer demand is strongest for egg, a new Internet banking entrant. egg is a subsidiary of Prudential, a very large and well-known financial institution, and could be argued to have allayed consumer perceptions of risk. While it is true that the link between Prudential and egg is not very prominent on egg Web site, present customers may be aware of the link through the considerable publicity that egg has generated since its launch. Hence, it may be fair to tentatively conclude that there is some evidence to support the concept that the association with an established parent brand, i.e. an umbrella brand, positively influences consumer purchase intentions.

8.3.5.3 Time Elapsed since Brand Launch

Finally, under branding it was also hypothesised that,

H 7: Purchase intentions are influenced by the strength of a brand, as proxied by the time elapsed since brand launch.

The time elapsed since brand launch is hypothesised to have an influence on purchase intentions. The rationale is that the longer the Internet banking service has been in operation, the more people will recognise, trust, and use the service. To evaluate this concept, the following dates are important. Firstly, date of domain name registration with bodies such as the InterNIC and RIPE is recorded. InterNIC is an organisation under the aegis of the U.S. Department of Commerce, while RIPE is a collaborative organisation open to all parties interested in the wide area IP networks. Both are non-profit organisations and have been offering registration services from almost the inception of commercial use Internet. It is in the interests of any organisation to get their domain names registered with these organisations. Secondly the date of the launch of transactional Internet banking service is recorded. These results are shown on Table (8.9).

	Title/Measure	BOS	BankNet	Barclays	Citibank	Co-operative	egg	First Direct	first-e	Halifax	Lloyds TSB	Nationwide	Natwest	NPBS	RBS	smile
1	Date of Web site registration	Nov-98	N/A	Nov-93	Feb-91	Oct-97	Oct-98	N/A	Apr-99	N/A	Jun-97	Mar-98	Nov-97	N/A	Sep-98	N/A
2	Launch of transactional Internet banking service	Nov-99	Jan-95	Jun-98	Apr-97	Mar-98	Nov-98	Nov-99	Sep-99	Oct-99	Apr-99	May-97	Nov-99	Jun-98	Jun-97	Oct-99

Table (8.9): Time elapsed since launch

It can be seen that Citibank has had the longest Internet presence. This was expected since Citibank is a large American bank. The Internet had its commercial beginnings in the USA, and commercial entities in the USA were the first to have an Internet presence. Barclays has had an Internet presence since 1993, three others have registered their domain names in 1997 and the rest registering their domain names after that. Five banks have chosen not to register their domain names with these bodies. The launch of transactional banking service on the Internet shows that Bank-Net has had the longest active use of the service, with First Direct and NatWest commencing their operations in November 1999.

There are a number of apparent anomalies in Table (8.9). Some banks appear to have launched a transactional Internet banking service prior to registering their Web sites. The only rational explanation would be that these banks have delayed the registration. The important variable for consideration is the launch date of the transactional Internet banking service.

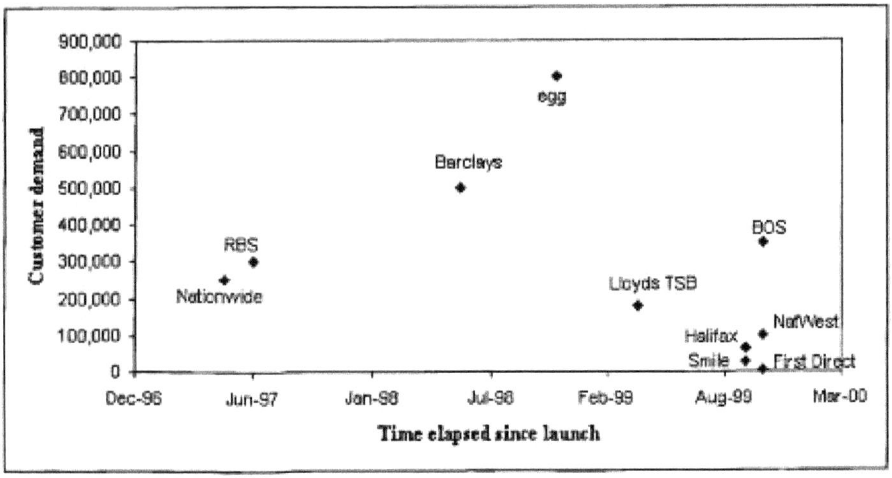

Figure (8.7): Time elapsed since launch and customer demand

The association between time elapsed since launch of Internet banking service and customer demand is demonstrated in Figure (8.7). The horizontal axis is arranged so that those with the least time lapsed will be away from the origin, and towards the origin would signify a longer lapse of time. If time lapsed influence purchase intentions, it would be indicated with the banks with a longer time lapsed since launch having a larger customer base. While the evidence is not totally conclusive at first glance, there are several important facts that require attention.

Although the launch date of BOS is November 1999, 350,000 customers use the service. This is because, prior to launching an Internet banking service, BOS operated a PC banking service. Similarly, both Lloyds and RBS previously operated PC banking services, hence their launch of remote delivery of electronic banking services in a broader sense is not as 'new' as it appears. Upon launching of their respective Internet banking services these banks actively encouraged their customers to convert the PC banking accounts to Internet banking service accounts. Perhaps following contributions on a discussion forum show the perception of some consumers.

> *".......the fact that smile is part of coop is good news. But must wait for some time before I bank with them - they are bound to have problems at the beginning, with time they will either iron out these and become really popular, or perish .. then I will decide with whom to bank........."*
>
> (gabriel lee, iii)

and,

> *".......... I have considered using First-e but basically don't trust them yet - no known brand, unkown quantity.."*
>
> (Philip Butterworth, 'iii')

Hence, on balance, it may be concluded that there is evidence to support the hypothesis that time elapsed since launch may influence purchase intentions.

8.4 Experience Evaluation of an Internet Banking Service

Once ex ante information is searched, acquired and evaluated, a consumer can purchase Internet banking service products. It was explained earlier that the unique qualities of Internet banking services enable the evaluation of the service to take place prior to purchase. To recap, this is because the demo facilitates a process where the consumer can experience both purchase and account operation procedures through simulation. It then follows that consumers gain experience qualities, ex ante, although such experience qualities in conventional services may only be acquired ex post. Therefore, it was hypothesised that consumer purchases are influenced by service quality, which can be evaluated using a number of interrelated measures.

8.4.1 Service Quality in Internet Banking

It was explained in section (4.5) that, (1) the tangible dimensions of a terrestrial banking service are the physical facilities, equipment and appearance of bank personnel, but in Internet banking, the equivalent tangible dimensions are Web site attributes, (2) responsiveness, assurance, customisation, involvement, and empathy dimensions of an Internet banking services are demonstrated by the provision of 'consumer friendly' opportunities for interaction and communication, (3) the duration dimension is demonstrated by the performance of a Web site. Hence service quality can be assessed by a collection of Internet scorecards that evaluate Web page characteristics, performance of the Web site and communication methods and features.

8.4.1.1 Web Site Characteristics

Apart from the important function of displaying Web site content, Internet banking service Web site characteristics that encompass navigation and design attributes portray an image, akin to the physical facilities found in a terrestrial banking organisation.

Therefore, it was hypothesised that,

H 8: Purchase intentions, mediated by perceived Internet service quality, are influenced by Web site characteristics.

In addition to the obvious aspects of colours, graphics and layout, aspects such as the use of frames and tables, likelihood of distortions under a multitude of screen specifications and settings are indicators, as shown in Table (8.10, next page), are evaluated.

Most Web sites state the optimum screen resolution that is best suited to view their Web site, where it was not stated, the resolution did not make an appreciable difference to the appearance of the Web site. BankNet's Web site is devoid of any graphics or colours, with black text on white background, hence this Web site was clear with no distortion whatsoever in any screen resolution. smile's Web site on the other hand is very colourful with frames and text arranged in a tabular format, and if the user does not alter the screen resolution to the recommended specification they may not observe the Web site as it is intended to be. A colourful Web site with a large amount of graphics does not necessarily consume more memory. The use of low definition simple graphics in 'jpeg' format can significantly reduce the memory requirement and speed up the download process.

	Title/Measure	BOS	BankNet	Barclays	Citibank	Co-operative	egg	First Direct	first-e	Halifax	Lloyds TSB	Nationwide	Natwest	NPBS	RBS	smile
	Ease of use/Navigation															
1	Entry guidance	7.0	4.5	7.0	7.5	7.5	5.0	7.0	8.0	7.0	6.5	6.5	5.0	7.0	7.0	8.0
2	Number of links from other web sites (Alexa count)	405	4,305	3	6,270	372	128	60	No data	532	142	555	280	41	630	No data
3	Ease of use	6.0	5.0	7.5	8.0	7.0	6.5	7.0	8.5	7.0	5.0	7.5	6.5	7.5	6.5	8.5
4	Website structure	6.0	3.5	7.0	8.5	6.0	6.0	6.5	8.0	7.5	5.5	6.0	6.0	7.5	5.0	8.5
5	Navigation consistency/ site index	5.0	3.0	7.0	8.0	6.0	7.0	7.0	8.5	8.0	4.5	7.5	5.0	7.5	5.0	8.5
6	Usefulness of hyperlinks	5.5	2.0	8.0	7.0	7.0	7.5	7.0	8.0	7.5	6.0	6.5	7.0	7.5	6.5	8.0
7	Hyperlink connotations	5.5	5.0	7.5	7.5	7.5	7.0	7.5	7.5	7.0	6.0	7.0	6.5	7.0	7.0	8.0
8	Hyperlink encoding	7.0	6.0	7.5	7.0	7.0	7.0	7.0	8.0	7.0	5.5	6.5	6.5	7.0	6.5	8.0
9	Search engine/facility	3.5	2.0	4.0	4.5	5.0	7.5	7.5	7.0	8.5	7.0	6.5	6.5	6.0	4.5	7.5
10	Homepage hyperlinks	8.0	4.5	5.5	8.0	7.0	6.5	5.0	8.5	7.5	4.5	7.0	7.0	7.5	7.0	8.0
11	New browser window	5.0	2.5	7.5	5.5	4.0	3.5	5.0	8.5	7.0	6.0	5.5	7.0	7.0	6.0	8.0
	Design															
1	Layout	6.0	3.5	5.5	7.0	5.0	6.0	7.0	8.0	6.0	6.0	5.5	6.5	6.5	6.5	8.0
2	Intelligent use of frames?	6.5	6.0	7.0	7.0	7.0	7.0	7.0	8.5	7.0	6.0	4.5	6.0	7.0	7.0	7.5
3	Intelligent use of tables	4.0	3.5	4.0	8.0	7.5	7.0	6.5	9.0	7.0	7.5	4.5	6.5	7.0	7.0	7.0

4	Use of colours	6.0	1.5	7.0	7.0	5.0	7.5	6.5	8.5	7.0	7.5	5.0	6.5	7.5	7.5	7.5
5	Use of graphics	5.5	2.0	7.0	7.5	4.0	7.0	6.5	9.0	7.0	7.5	5.0	7.0	6.5	7.5	7.0
6	Likelihood of distortions	7.5	8.5	7.5	7.0	7.0	7.0	7.5	7.5	7.5	7.0	6.5	6.0	7.0	7.5	7.0
	Others															
1	Online help features	7.0	7.0	7.0	7.0	7.0	7.0	7.0	7.0	7.0	7.0	7.0	7.0	7.0	7.0	7.0
2	Interactivity	5.0	5.0	5.0	5.0	5.0	5.0	5.0	5.0	5.0	5.0	5.0	5.0	5.0	5.0	5.0
3	Language customisation	1.5	1.5	1.5	1.5	1.5	1.5	1.5	1.5	1.5	1.5	1.5	1.5	1.5	1.5	1.5
4	Brower suitability function	n	n	n	y	n	n	n	n	n	n	n	n	n	n	n
	Average score	5.7	4.0	6.3	8.8	5.9	6.3	8.4	7.8	6.8	5.9	5.8	8.1	6.7	8.2	7.3

Table (8.10): Website characteristics

Smile and first-e have better hyperlink connotations, while BOS and BankNet are comparatively poor. They (smile and first-e) have succeeded in producing comparatively better-encoded hyperlinks. When a consumer 'clicks', on a hyperlink they should take the user to the intended location or execute the task it is intended to. Here again, with the exception of BankNet, all other Web sites contain hyperlinks that perform their intended task. BankNet contains some hyperlinks that result in a default error page in the browser. first-e and smile have the most user friendly and logical navigational consistency, thereby making it easy for a consumer to arrive at the appropriate destination within the Web site. These sites appear to be consumer friendly because during the navigation process the consumer is able understand the Web site structure, thereby navigate more efficiently. The primary aid within these sites that makes this process possible is the ever-present Web site directory structure, which is very prominently displayed.

Another very useful feature that is found in the first-e Web site that expedites this process, is a series of drop down menus. A site index is a useful feature that aids in this process and so is the automatic opening of new browser windows; for example, if a consumer commences browsing on the 'job prospects' link, while they were on the Internet bank Web site, it is logical that a new browser window opens, without closing the bank browser window: this is an example of segregating browsing by subject areas. smile and first-e, have specifically designed their Web sites to follow this subject-oriented method of browsing, thereby making sure that the consumer can return to the bank Web site with the minimum effort. There is another equally useful if not more important function that a new browser window fulfils. It is a security feature. A new copy of the browser will only accept and send coded messages to and from the Bank Internet server. Some Web sites, such as Halifax, contain a dedicated site search engine that further refines the search process. However, Internet bank Web sites have tended to avoid search engines because of the specific nature of the output that search engines produce. Most search engines, search Web page content by looking for their Meta tags, and hence there is a danger that even though the Web page is not very relevant to the user query, the search engine might select it as the best page match, if the Meta tags have been selected in an appropriate way. Meta tags are a method of identifying the contents of a Web page by the use of keywords in HTML code.

Considerable attention is paid to the evaluation of Web site format. In this regard, the two most popular methods of organising Web pages, i.e. frames and tabular or column format are important considerations. In order to enhance the design, it is common to utilise frames in Web pages, this has advantages and disadvantages. On the negative side,

- on some older versions of browsers frames may be displayed m a distorted manner,
- they consume more computer memory,
- they take a longer period to download, if the computer screen has been set at a different resolution to the frame design, users will get a distorted view, and
- more prone to errors if the Internet connection is not stable or fast.

Popular browser upgrades are easily and freely available, but upgrading is not a simple option for users with older computers. However, it is acknowledged that these problems are becoming less of a concern as more Internet users begin to use newer and faster machines. One problem that is difficult to mitigate is the complexity of making bookmarks for Web sites with frames. On the positive side, efficient use of frames is very useful for navigation. An almost ubiquitous use of frames is using the left-hand side of the Web page frame for the site index or subject guide. When a user scrolls down a page, the frame stays stationary, providing all the necessary links at all times, irrespective of the location on site. A number of Web sites do not use frames altogether, while at the other end of the scale, Nationwide's Web pages comprise of two frames per page, occupying the header and the footer. There is a fierce debate about the use of frames among Web authors and the benefits or dis-benefits of frames are non conclusive. Tabular presentation of content is a popular alternative method of presenting text. Where tables are used, the table borders are usually are not present, however the primary objective of utilising tables is to express the text width as a proportion of the screen size. By controlling the width of text fill area on the monitor, it is possible to create a Web page that is easy to read with restricted horizontal text length non-dependant on screen resolution. In this regard first-e and Citibank are the Web sites of noteworthy presentation.

Other attributes of a Web site that contribute to service quality of an Internet banking service are interactive features, browser suitability functions and online help features.

The interactivity of a Web site relates to the optimisation of hypermedia attributes by providing custom made solutions. In other words, this allows users to customise preferences, for example the content that users prefer on the login page. It also enables the bank to provide bespoke solutions online for users who have provided appropriate information. None of the fifteen banks appear to have made any notable progress is this regard. smile and first-e have attempted in some way to make the interface appear personalised, however these efforts fall far short of what can be achieved. There is another aspect to interactivity of Web sites, i.e. the provision of interactive modules on Web sites. It is technically feasible to provide interactive loan calculators, exchange rate converters, mortgage calculators, etc. on Web sites. The provision of these facilities should work to the benefit of the banks in several ways. Firstly, the customers could judge for themselves their commitments for something like a loan, which might otherwise require the assistance of bank staff, thus relieving them and thereby representing a resource saving to the banks. Secondly, these facilities can draw new customers into the bank, for consumers perceive that they are in control of their account, which as stated elsewhere is one of the primary attractions of Internet banking services. Thirdly, as noted earlier, an Internet banking site has an opportunity to position itself as the focal point of electronic commerce. Interactive functions could be a focal point to attract shoppers. The number of banks that provide these facilities is in the minority, with only NatWest and egg providing online calculators to calculate interest accruals on savings accounts and interest payments on borrowings.

Many of the Internet bank accounts are designed to operate on certain minimum computer specifications, with regard to security features, applications, etc. These specifications are usually expressed by specific versions of browsers, and it is useful to help consumers with this assessment process. Therefore browser suitability function is a small application programme within the Web site that communicates with consumer's computer and decides whether or not the browser is compatible. Only Citibank provided this function, other banks simply stated what consumers should have in order to bank with them. In today's environment a 'help' feature is ubiquitous in almost every computer related application. Internet banking services is also an application, and moreover it is relatively new, thereby necessitating the presence of an online help feature. Most banks provide such a feature, with the exception of Citibank and BankNet. Help functions by first-e, smile and Halifax are very useful.

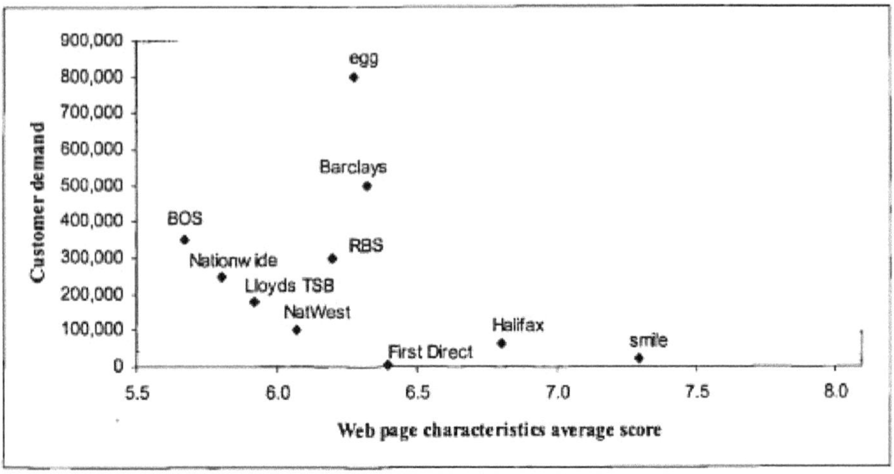

Figure (8.8): Web site characteristics and customer demand

Figure (8.8) shows a rudimentary assessment of a hypothesised variable, Web page characteristics, and its relationship with customer demand. The un-weighted average score of Web site characteristics is given along the horizontal axis. A positive relationship would be signified by the best Web page characteristics attracting the most number of customers. There is no evidence of a simple positive relationship. Detailed analysis is hampered for reasons reiterated earlier. While no simple relationship is visible, it can be seen that Barclays and egg have been successful in attracting a large number of customers in spite of having mode Web page characteristics. Perhaps their success is due to the time lapsed since the launch of their transactional Internet banking service. However Nationwide and RBS have had similar length of time of operational Internet banking service accounts, but these have not been translated into higher customer numbers. BOS in spite of having the poorest evaluated Web site characteristics, have the third highest customer demand. Therefore a tentative conclusion that can be drawn is that there is little evidence to conclude that there is a clear relationship between Web site characteristics and consumer purchase intentions.

8.4.1.2 Performance Aspects

It was argued that, the performance of a Web site and related components influence service quality. Therefore, it was hypothesised that,

> ***H 9: Purchase intentions, mediated by perceived Internet service quality, are influenced by the performance of the Web site and related components.***

Admittedly, performance is an all embracing term, it encompasses, the performance characteristics of the Web site, the downloading speeds of the Web site and the performance of the demo component. Accordingly the evaluation is carried out under three headings. It is important to recognise that the influence of the demo component in the information search and acquisition has already been discussed under section (8.3.1).

Performance Characteristics of the Web Site

It can be seen from Table (8.11, next page) six banks require the user's computer to contain additional software such as ActiveX controls and plug-ins for optimum performance to merely browse through their Web sites. When a computer is bought, it is not usually preloaded with such software programmes, therefore this means that in most cases consumers have to download required software from the Internet.

While these software programmes may be freely downloadable, not all consumers may wish to do so. Firstly, such downloading consumes time, and hence it is a time related cost in the UK Internet access environment that existed in January 2000. Secondly, additional software generally slows down the computer. Thirdly, if a user's browser is set to medium or higher security levels, it is not safe run an ActiveX controls and a new window pops up alerting the user that they may be compromising the security levels by running such applications. Downloading software from the Internet always carries the risk of importing viruses to the computer. In these situations it may be prudent for the bank Web site to offer an option of bypassing such controls and offering the user more choice in browsing the site, albeit without some animation features but with essential content. NatWest and Halifax offer this option, with Halifax going a step further in offering a text only version of the Web site devoid of any graphics.

	Title/Measure	BOS	BankNet	Barclays	Citibank	Co-operative	egg	First Direct	first-e	Halifax	Lloyds TSB	Nationwide	Natwest	NPBS	RBS	smile
	Operation															
1	Animation/multimedia plug in requirements	v	n	n	n	n	n	v	v	v	n	n	v	n	n	N
2	Any additional software to download?	v	v	n	n	v	n	v	v	n	n	n	n	n	n	y
3	If so, uninstall facility?	v	n	n/a	n/a	n	n/a	n	v	n/a	n/a	n/a	n/a	n/a	n/a	N
4	Information retention in computer	n	n	v	n	n	n	n	v	n	n	n	n	n	n	N
5	Off line operation	n	n	v	v	v	v	n	v	v	v	v	y	v	v	y
6	Cookies	v	n	v	v	n	v	v	v	v	v	v	v	n	v	y
	Access to Web site/operation															
1	Web congestion	7.5	4.5	7	7.5	7.5	7	7.5	7	7.5	7	6	7	7.5	7.5	7.5
2	"Bugs" in the operation	7	3	7	8	7	7	7.5	7	7.5	2.5	7	6.5	7.5	7.5	7.5
3	Planned maintenance of Web site (hrs per week)	37	0	0	0	9	0	0	0	0	28	0	0	0	0	0

Table (8.11): Performance characteristics of the Web site

In addition to generic multimedia software requirements of ActiveX controls and plug-ins, six banks require the user to download account specific bespoke Java applets, except in the case of BankNet, where the download is non-Java. All these Java downloads will reside in the consumer's computer hard disk. The rationale for favouring Java is that, it is believed that in future a majority of consumers would want to access their Internet bank accounts through a variety of devices and not limit themselves to a computer. Possible devices include, but are not limited to, mobile phones, personal assistants, electronic organisers, etc. Java is a language that is very versatile and already some of these devices are capable of supporting Java, hence the choice. Additionally, as the processing capacity of electronic devices increase, they should be able to process Java programmes much more quickly.

On the downside shortcomings of Java are,

1. At present all of these Java based offerings are designed to communicate with the bank server from the moment they are started. Therefore, it is not possible to commence the banking process in offline mode and get to the login stage and go online, as may be possible with HTML based offerings.
2. It follows from the above that it would consume comparatively more time for the login process when compared to browser-based offerings, and this leads to higher costs in terms of telephone charges.
3. If a customer attempts to access the account from an office computer that is part of a Local Area Network (LAN) or a Wide Area Network (WAN) a corporate firewall may prevent the downloading of executable software programs from the Internet.
4. If a customer were to use different computers for access on different occasions, the downloading process is repeated and there is a possibility that sensitive information may be captured and retained in the computer's memory without the customer's knowledge. Third parties may access such information by appropriate means.
5. This approach may be functionally a wise choice, but leaves a small number of potential customers with older machines that lack a Java compatible browser (i.e. older releases of browsers running on lower specification computers) unable to use the service. Popular browser upgrades are easily and freely available, but upgrading may not be an option for users of older computers with slow processors.

6. Java applets do not run smoothly on versions of Windows operating systems prior to Windows 95. However, it is acknowledged that this is becoming less of a problem as more and more Internet users begin to use newer and faster machines.

A feature that merits an evaluation is whether it is possible to uninstall any of these Java components once they are installed, if the user wishes to do so. Only two banks, BOS and first-e, offer an uninstall option.

There are a number of disadvantages of having to install additional software, and the following contribution effectively summarise a major consumer concern.

> *"For a perfect example of how not to launch a service look at first-e. They require that you install several browser addons before you can use that service (try that next time you are in a far flung part of the world connected to the Internet via a few pieces of string .... "*
>
> (chriseb, Motley Fool)

Some versions of browsers, such as Internet Explorer (IE) version 5 and above, offer the option of synchronising a Web site. To 'synchronise' is to make an entire Web site available offline. Once the Web pages have been synchronised a user can browse the site when the computer is not connected to the Internet. For example, a user can view Web pages on their computer when they do not have a network or Internet connection. Or they might want to read Web pages without tying up a telephone line at a specific time. The synchronizing process allows the user to specify the amount of content to be made offline, this is usually expressed as a single page or a page and its links up to a specified depth. The user can also specify the computer to update automatically, should the Web page content change. Further, there are other options that are available when synchronizing a Web page, from just saving the text, to saving all of the images and text needed to display that page as it appears on the Web. Apart from three banks, BOS, BankNet and First Direct, all other banks allowed user to synchronise Web pages and view in offline mode.

One of the major primary features of an Internet banking service is the ability to carry out banking transactions devoid of time constraints. In practise this means that consumers expect an Internet banking service to function, 24 hours a day and 7 days a week, popularly referred to as '24x7'. Three banks have scheduled maintenance programmes, ranging from BOS that is inaccessible for 37 hours in a week to Cooperative bank, which is closed for 9 hours per week. All others are operational on a 24x7 basis.

Those banks that close their Web sites appear to be operating against one of the very objectives of Internet banking.

The majority of Internet bank Web sites require the user's computer to accept cookies. Cookies contain information about the user and their preferences. It can contain a large portion of their browsing pattern, their purchase behaviour, etc. or it might only contain a record of which pages within the site the user visited. The purpose is to help the Web site customize the view for the user the next time they visit. Only the information that users provide, or the choices they make while visiting a Web site, can be stored in a cookie. For example, the site cannot determine the user's e-mail name unless they choose to type it. Allowing a Web site to create a cookie does not give that or any other site access to the rest of the user computer, and only the site that created the cookie can read it. Nevertheless some consumers may perceive this as an invasion of browsing privacy. Both types of browsers, IE and Netscape is set up by default to allow the creation of cookies; however, one can specify that they be prompted before a site puts a cookie on the hard disk. Some consumers may not be aware of these options in browsing and thus may allow these banks to collect information about them without their knowledge.

It appears that banks with simple designs, such as Barclays, egg and Nationwide, appeal to a large number of customers. Customer demand appears to be comparatively lower for banks Web sites with additional software download requirements.

Web Site Speed

Web site downloading speed is part and parcel of the performance of the Web site. Many commentators have observed that, in general speed of downloading has increased in recent times (Jayawardhena and Foley, 2000). It is important to recognise that speed is also dependent on the user's computing hardware and method of connection. However a site owner (i.e. the banks), can speed up the downloading process by using low memory graphics in their Web pages and hosting the Web sites in high capacity servers.

In order to get a more realistic measure of Web site downloading speeds, and to mimic the usage patterns of a customer, all fifteen Web sites were repeatedly accessed, on consecutive days. All Web sites recorded a reduction in download times over time, and recorded in Table (8.12). This is because, most of the newer browser software, including IE, saves most of the graphical components of a site in the computer memory by default. Therefore, when a consumer access a Web site on more than one occasion, the computer communicates with the host computer to ascertain which saved images can be used, if a match is found it does not download those from the host thus reducing download times of a recently downloaded site. Barclays and BOS in particular were very slow to download. Consumers using older hardware with older versions of browsers may find that download times take even longer.

	Title/Measure	BOS	BankNet	Barclays	Citibank	Co-operative	egg	First Direct	first-e	Halifax	Lloyds TSB	Nationwide	Natwest	NPBS	RBS	smile
	Web site speed															
1	Downloading time	5.0	6.0	4.5	6.5	7.0	7.5	7.5	7.0	7.0	7.0	7.0	7.0	7.5	7.5	7.5
2	Offline operation	n/a	n/a	5.5	6.0	7.0	8.0	n/a	8.0	4.5	7.5	7.0	7.5	7.5	7.5	7.5
3	Access speed (in kb/s)	No data	14.4>	No data	28.8>	14.4>	14.4>	14.4>	No data	No data	No data	No data	14.4>	14.4>	14.4>	No data
	Notes Co-operative, First-e, smile - demos run as java applications Some sections of the BankNet account needs bespoke software application lo operate, but the demo runs on the browser.															

Table (8.12): Web site speed

Only two Web sites, Citibank and Cooperative, give an indication of approximate download and account operation times for average usage. Download times were not calculated in seconds, rather they were evaluated along the ordinal scale. This is because, for reasons outlined earlier the downloading speed is a function of many factors, and attempting to measure it in seconds cannot be justified to be an objective evaluation. A more robust measure of download speeds is captured by the Alexa access speeds. This measure does not suffer from individual user's computer specification related issues. It is measured in kb/s (kilobytes per second). This refers to the Web site's effective transfer rate, calculated during Alexa agents' periodic crawls of the Internet. Only Citibank allowed a transfer rate of between 14.4 kb/s to 28 kb/s, and where data were available, all other banks allowed a transfer rate of less than 14.4 kb/s. Additionally, most of the sites utilise Java Applets in order to provide optimal transactional facilities. An Applet is a program written in Java (a programming language) to run within a Java compatible Web browser. The merits and shortcomings of Java were discussed in the above section. In general it can be concluded that access speed are higher where Web designs are simple, devoid of complex frames, etc.

Demonstration Performance

This is the third aspect of Web site performance evaluation. As shown in Table (8.13) an important factor that can be evaluated by analysing the demo is the platform freedom offered by the Internet banking services for account operation. It appears that this is a concern for many consumers who are considering a purchase of Internet banking services. The following are typical contributions.

> *"..., i want to be able to use a variety of platforms,. Don't these people ever use their own systems?......"*
>
> (chriseb, Motley Fool)
>
> *".Does any nb any online banking service work with apple-mac?.."*
>
> (VIVIAN HORWITZ, iii)

	Title/Measure	BOS	BankNet	Barclays	Citibank	Co-operative	egg	First Direct	first-e	Halifax	Lloyds TSB	Nationwide	Natwest	NPBS	RBS	smile
	Demo Operation															
1	Animation/multimedia plug in requirements	y	n	n	n	n	n	Y	n	n	n	n	n	n	n	y
2	Additional software	y	y	n	n	y	n	N	y	n	n	n	n	n	n	y
3	If so, uninstall facility?	y	n	n/a	n/a	n	n/a	N	y	n/a	Nia	n/a	n/a	n/a	n/a	n
4	Off line operation	y	y	y	y	n	n	N	n	y	y	y	y	y	y	n
5	Platform freedom	*n*	*y*	*y*	*y*	*y*	*y*	Y	*n*	*n*	*y*	*y*	*y*	*y*	*n*	*n*
	Demo Download Speed															
1	Downloading lime (in minutes)	12	*			3			10							12

Notes

egg has no demo

Co-operative, First-e. smile - demos run as java applications

Some sections of the BankNet account needs bespoke software application to operate, but the demo runs on the browser.

* - It was not possible to download all components of the BankNet demo

Table (8.13): Demo performance

Most home computers users would use computers running either on Windows or Apple Macintosh operating systems. All banks except BOS, first-e, Halifax, RBS and smile, offer platform freedom between these two operating systems, and the others only operate on a Windows operating system. The root of this lack of platform freedom is the incompatibilities that exist between these two popular operating software programmes. Considering the fact that about 15% of the computers in this country are operated on non-Windows environments this does mean a significant proportion of potential customers are left out. Of these, RBS will not operate even on UNIX run computers while others will. UNIX is a computer operating system that is favoured by a large number of Internet operations, and not limited to banking. It is considered to be comparatively stable, especially compared to Windows, and offers many security features that other operating systems cannot offer. Therefore it is indeed rather surprising to note that RBS will not operate in such an environment. However it is not user-friendly like the other two popular operating systems, and for this reason very few individuals would use UNIX as their operating system.

Four demos, those of BOS, Co-operative, first-e and smile, need downloading as a separate file for operation. Three of these take 10 or minutes to download, which is a significant amount of time.

Performance aspects of a web site have been evaluated under three headings. Eye balling Tables (8.11), (8.12) and (8.13) indicate that there is no obvious relationship between Web site performance and customer demand. However a number of forum contributions show that consumer purchase decisions may be influenced by the performance. Consider the following,

> *"..Anyone managed to set up an account with Smile yet?.... Early days I know - but their system performance does not bode well. Website seems to be overloaded...."*
>
> (hanyson, Motley Fool)

Therefore, all things considered it is tentatively concluded that there is evidence to support the hypothesis that Web site performance may influence consumer purchase intentions.

8.4.1.3 Communication Methods

It was established that aspects of assurance, empathy, credibility and communication are service quality dimensions that are demanding to evaluate in an Internet banking service. Consequently these dimensions of service quality are measured through a combination of variables. These include the evaluation of speed, quality and security features of communication.

Therefore, it was hypothesised that,

> ***H 10: Purchase intentions, mediated by perceived Internet service quality, are influenced by the responsiveness of communication, speed and accuracy of responses to customers.***

In the absence of face-to-face contact, aspects of speed, accuracy and responsiveness of communication through other means become of paramount importance. Other reasons for the apparent need for a communication medium are discussed in the previous chapters. Surprisingly, Barclays, Citibank and NatWest did not have an email link or a general Web query form to make specific inquiries, five banks had an email link, six had Web based forms, and Nationwide gave the consumer the opportunity of contacting the bank through either of the two methods. A standard query, as shown in section (7.3.3), was sent to these banks. Table (8.14) captures characteristics of the communication interface.

In addition to the traditional communication channels such as post and telephone, all except three banks offer either an email-based or Web-based communication channel for consumers. Of importance is to indicate whether such communication can be made over a secure server. In the case of three banks that did not offer consumers an Internet based method of communication, an email is sent to the 'postmaster' of the respective bank.

	Title/Measure	BOS	BankNet	Barclays	Citibank	Co-operative	egg	First Direct	first-e	Halifax	Lloyds TSB	Nationwide	Natwest	NPBS	RBS	smile
	General															
1	Availability	y	y	n, sent to p	n, sent lo D	y	y	y	y	y	y	y	n	y	y	y
2	Type	w	e	e	e	e	w	w	e	w	w	w+e	e	w	w	e
3	Secure?	n	n	n	n	n	y	n	n	y	n	n	n	n	n	n
4	Bug free	8.0	n/a	n/a	N/a	n/a	7.5	8.0	n/a	8.0	8.5	8.0	n/a	8.0	7.5	n/a
5	Ambiguity	6.0	6.5	n/a	Nia	6.0	7.5	8.0	7.5	7.5	7.5	7.5	7.0	6.0	7.0	7.5
	Other Features															
1	News Distribution	5.5	3.5	6.5	6.5	8.0	8.0	9.0	9.0	6.0	5.5	5.0	5.0	5.0	5.5	9.0

Notes

Where there is no Web interface/program for communication, an email number is gives for communication. In which case 'software bugs' do not arise. w - Web based; e - email; w+e - Web based and email

n, sent to p - communication interface not available, an email sent to the postmaster

Table (8.14): Communication methods performance characteristics

All replies received are analysed to gauge the speed, reliability, responsiveness and accuracy. Table (8.15, next page) records all these aspects. Citibank did not reply, while Barclays sent an automated acknowledgement of receipt, which is instantaneous, but no specific reply to the email. Six banks replied within a day, four others replied in under three days, NPBS took almost a week to reply. Such delays may have a negative influence on purchase intentions. Consumers may perceive that if a problem occurs, once they become customers, similar delays may be expected in attending to problems. Crucially, reminders or repeat emails were not sent, since a consumer would note a non response as, (1) an indication of the level of service they would receive once they become a customer, should they wish to do so, and (2) as an indicator of how eager the bank is to make them customers. By and large the responses are useful, they show that the banks understood the issues, showed sympathy with the consumer and attempted to show that their concern is consumer interest. The reply from BankNet addressed only one issue and appeared not to have understood the enquiry, i.e. if the bank cannot understand a consumer query and respond accordingly, what is the competence of the bank and what sort of service can be expected? Consumers might perceive that the service is not likely to function smoothly and may expect many difficulties in the operation of the account.

	Title/ Measure	BOS	BankNet	Barclays	Citibank	Co-operative	egg	First Direct	first-e	Halifax	Lloyds TSB	Nationwide	Natwest	NPBS	RBS	smile
	Speed and accuracy of response															
1	Email/ query sent on	1707, 07/02	1742, 07/02	0644, 08/02	0621, 08/02	0606, 08/02	0524, 08/02	2221, 07/02	0639, 08/02	0647, 03/02	0748, 03/02	1222, 03/02	0612, 08/02	0628, 08/02	0631, 08/02	0635, 08/02
2	Reply Received	1301, 08/02	1821, 07/02	None	None	1941, 09/02	1454, 09/02	1029, 08/02	1629, 08/02	0853, 03/02	1028, 04/02	1350, 03/02	1540, 09/02	1205, 15102	1418, 10102	0908, 09102
3	Re-sponse time (in hrs)	19.5	0.5	n/a	n/a	37.5	33	12	9	2	15	1	33	174	56	26.5
	Responsiveness of Communication															
1	Useful response	6.0	3.5	n/a	n/a	9.5	8.0	7.5	8.5	6.5	6.5	9.0	8.0	9.0	6.5	9.0
2	Under-standing	7.5	3.0	n/a	n/a	9.5	9.0	8.0	8.0	7.5	7.5	9.0	7.0	9.0	6.0	9.0
3	Sympa-thy	6.5	4.0	n/a	n/a	9.0	8.5	8.0	8.0	8.0	7.5	9.0	8.0	9.0	5.5	9.0

4	Customer interest	7.0	3.0	n/a	n/a	9.0	9.0	8.0	9.0	6.0	7.5	9.5	5.5	9.0	6.5	7.5
	Competence															
1	Competence and adequacy of response	6.5	3.0	n/a	n/a	9.5	7.5	7.5	8.5	6.0	6.5	9.0	8.0	8.5	6.0	9.0
2	Error free	8.0	6.5	n/a	n/a	9.0	9.0	9.0	9.0	8.0	7.5	9.0	8.0	8.0	8.5	8.0
	Other Features															
1	News Distribution	5.5	3.5	6.5	6.5	8.0	8.0	9.0	9.0	6.0	5.5	5.0	5.0	5.0	5.5	9.0
2	Ambiguity	6.0	6.5	n/a	n/a	6.0	7.5	8.0	7.5	7.5	7.5	7.5	7.0	6.0	7.0	7.5
3	Bug free	8.0	n/a	n/a	n/a	n/a	7.5	8.0	n/a	8.0	8.5	8.0	n/a	8.0	7.5	n/a
	Average score	6.8	4.1	6.5	6.5	8.7	8.2	8.1	8.4	7.1	7.2	8.3	7.1	7.9	6.6	8.5

Table (8.15): Responsiveness of communication

Figure (8.9, next page) demonstrates the relationship between customer demand and service quality dimensions as depicted by features of communication. Barclays is notably absent because they did not respond to the communication. A logical expectation in this instance is that a high score in responsiveness of communication will be rewarded with high customer demand. On visual examination, there appears to be no obvious indicative movement in the two variables. In spite of having communication systems that are inferior to others, RBS and BOS, have larger customer bases and appear out of tune with others. This is probably due to reasons stated earlier in this chapter, i.e. these two banks had a PC banking system prior to the launch of their Internet banking service. For reasons opposite to this, smile, a new entrant, has a small customer base in spite of superior communication features. However, egg on the other hand, has good communication features and has the largest customer base.

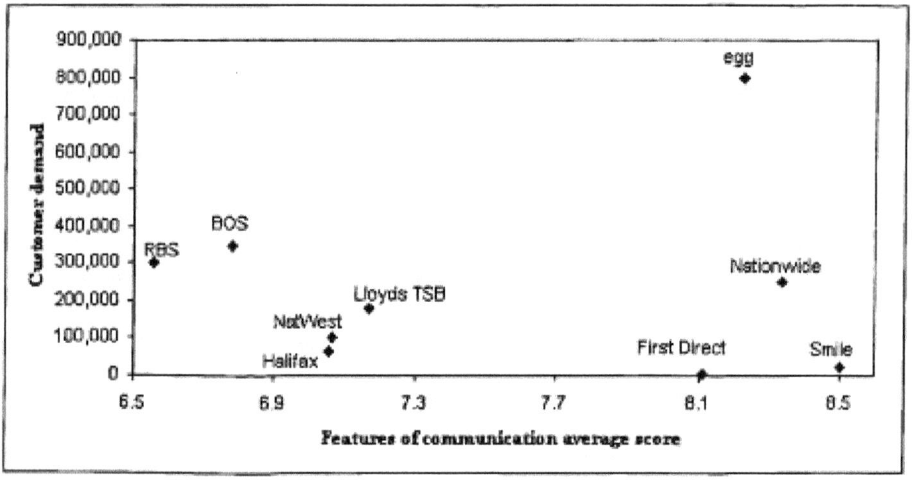

Figure (8.9): Features of communication and customer demand

Consider the following,

> *"........... 7 e-mails on consecutive business days and not a single response from Smile. Even an e-mail to their feedback address elicited no response. All I was asking was a simple question not addressed in their FAQ. Join Smile - you must be joking ......"*
>
> (Adrian Lee, iii)

This contribution show that some consumers may be served in a manner that is not suggested by the mystery-shopping exercise, but it nevertheless reinforces the general arguments in this section. Therefore it is suggested tentatively that purchase intentions are influenced by communication methods and related features.

8.4.2 Privacy and Security

A crucial element that is essential to the formation and maintenance of a banking relationship between a consumer and a bank is security (Jayawardhena, 1 999a and 1999b). An Internet bank, like any other bank, must be able to instil confidence in consumers by adequately informing the consumer of the measures in place to protect both confidential information and most importantly their assets. Therefore, it was hypothesised that,

H 11: Purchase intentions, mediated by perceived Internet service quality, are influenced by Website security.

Banks instil consumer confidence through the Web site and associated components, the evaluation of which is given in Table (8.16). The objective is to demonstrate to the consumer that they are being individually treated in an honest, speedy, secure and accurate manner. Additionally, attributes like brand (discussed above in section 8.3.4) assist in instilling confidence in consumers.

Many security features are employed in these Internet banking service operations. Encryption technologies form the backbone of these measures. In addition measures such as passwords, personal identification numbers (PIN), and other security codes are important.

	Title/Measure	BOS	BankNet	Barclays	Citibank	Co-operative	egg	First Direct	first-e	Halifax	Lloyds TSB	Nationwide	Natwest	NPBS	RBS	smile
	Security Measures Used															
1	Pages with secure connection	1.5	7	7.5	7	7	8	7.5	7	5	7	8	4	0	5	7
2	Encryption	no data	no data	no data	128	no data	128	no data	128	128	128	128	no data	no data	168	128
3	Password	v	v	v	y	y	y	y		y	y	y	y	y	y	y
4	Other unique customer identifiers (number)					5	3	2				2		1	2	5
5	Unique transaction numbers	n	n	n	n	n	n	n	v	n	n	n	n	n	n	n
6	General explanation	4.5	4	3	7	7.5	8.5	8	8.5	7	6	7.5	5.5	7	7.5	8
7	Overall security perception	5	5	6	8.5	7.5	8.5	7.5	8	7	6.5	7.5	6.5	6	7	8.5
8	Confidence	6.5	3	7.5	9.5	8	8.5	7.5	8	7	7.5	8	7.5	6	7.5	9
9	Lockout						y		y				y	y		y
10	Secure communication	n	v	v	y	v	v	v	v	v	v	v	v	n	n	v
11	BS for Security	n	n	n	n	n	n	n	n	n	n	n	n	n	n	y

	Language															
1	Explicit	4	4.5	2.5	7.5	8	8	7.5	8	7	5.5	8	6.5	7	7.5	8
2	'Small print'	3	5	4.5	7	6.5	7.5	7	8	6.5	5	7	6	7	7.5	8
	Regulatory															
1	Country of banking license	UK	UK	UK	UK	UK	UK	UK	France	UK	UK	UK	UK	UK	UK	UK

Table (8.16): Security features

It is fair to state that Internet banking services have been made possible only because of encryption technologies. Until a few years ago, the only way banks could guarantee security for customers was to make them dial into the bank's private network. The transactions are secure, but the cost of supporting a dial-up infrastructure makes it economically impractical, and acted as an entry barrier. However, the US government has approved the use of strong (128-bit) encryption for banks and other financial institutions outside the US, as long as it is used exclusively for integrity and authentication, and it can be clearly established that there is no need for the end user to employ longer keys for the general encryption of messages. This means that the banks providing such services are able to move their PC banking services to the Internet, and new entrants could commence Internet banking service operations.

Providing customer access to strong encryption for people outside the US without requiring them to employ the longer keys for general encryption has multiple technological solutions. One solution, server-gated cryptography (SGC), available with Microsoft's Internet Information Server, allows the standard Microsoft or Netscape browsers, with 40-bit encryption (all browsers in general use outside the US cannot acquire 128 bit versions), to talk to a server using 128-bit encryption. It utilises Securities Service Layers (SSL) with a modification: the server asks the customer's computer to switch to 128-bit encryption during the initial handshake sequence, and when the session is over, switches back to 40-bit encryption.

Name of bank	Technology	Encryption key length
Bank of Scotland (BOS)	SGC	128 bit
BankNet	SSL	40 bit
Barclays	SGC	128 bit
Citibank	SGC	128 bit
egg	SGC	128 bit
First Direct	SGC	128 bit
Halifax online	SGC	128 bit
Lloyds TSB online	SGC	128 bit
Nationwide	SGC	128 bit
NatWest	SGC	128 bit
NPBS Netmaster	SGC	128 bit
Co-operative	Java	128 bit
first-e	Java	128 bit

smile	Java	128 bit
RBS	ActiveX	168-bit triple data encryption standard

Table (8.17): Security technologies

As it can be seen from Table (8.17) that apart from SGC and basic SSL technology, some banks have used both Java and ActiveX technologies to achieve the same effect. By embedding 128-bit encryption in an applet that is downloaded to the customer's PC specifically to run the banking software, banks can deliver 128-bit encryption for their application while maintaining the 40-bit encryption for other browser-based applications. Banks using Java-based applications, such as the Co-operative Bank, enjoy an advantage over those like the Royal Bank of Scotland that use ActiveX applets. This is because Java applications work on a much wider range of platforms. Consumers may experience problems running ActiveX applets on non-Windows platforms, or even Windows browsers other than IE. Complying with statutory standards (for example, BS 7799 for Information Security Management) appeals to consumer security perceptions and helps to build confidence. Only smile complies with BS 7799. Although all banks are evaluated to be secure, the evaluation reveal that the manner in which the information is presented to the consumers is very poor and it is evident that adequate attention has not been made to get the message across to the consumer.

A number of unique pieces of information are further utilised to increase the level of security. All the banks require that the consumer key in between three to five different unique pieces of information to log into the account. The customer may decide some or all these unique pieces of information, or these may be generated by a computer in the bank through a process of random selection. For instance, egg relies on three customer generated pieces of information, while Nationwide customers have a single piece of this information that is self determined and three others generated by the bank. Remembering up to three bank generated pieces of information (which are series of numbers wi.th up to 10 digits) is by no means an easy task, and is likely to put off consumers.

Thought not widely used, assigning a unique transaction number per transaction is a further measure of security. This is a system known as "challenge response logic" and is carried over from telephone banking. In its simplest form, this involves a list of code numbers sent to every online customer and used in sequence, in combination with their password or personal identification number (PIN). This gives each transaction a unique code, it is only first-e that uses this system. Banks that rely on Java technology further enhance the level of security. The bank and the customer's browser will each generate a private key (which is generated when the customer is asked to move the mouse over the cursor), and exchange them to create a password between themselves. Every 20 minutes this password will change, meaning that a hacker would only have 20 minutes to break the password to compromise security. Such a task is inconceivable with the present level of technology. It is equally important to carry out communication over a secure server. Although it appears to be a mandatory requirement, three banks, BOS, NPBS and RBS do not provide consumers with an opportunity to communicate over a secure server.

While some consumers are unlikely to understand these intricacies of Internet security, a fair proportion of consumers are aware of the different technologies of providing security. Such issues are discussed in immense detail on some of the discussion boards, and would stretch onto several pages if reproduced here. some consumers may wish to access their Internet banking accounts through firewalls, and depending on the configuration this might present problems. The following is such a comment,

> *"...I want to be able to connect via a firewall, I want to be able to use 128-bit encryption. Don't these people ever use their own systems?......"*
>
> (chriseb, Motley Fool)

Of the banks reviewed here, first-e is the only bank to operate throughout the European Union -UK, France, Germany and Spain. Under the European Union rules it is possible for an organisation with a banking license from one of the member countries to operate throughout the union. The other exception is Citibank, which is a part of a large American bank, operating in the UK under a UK banking license. first-e is operating with a French banking license and based in Dublin, Ireland. Hence, first-e the Internet bank is not currently a member of the UK Banking Ombudsman Scheme or a signatory to the British Bankers Association (BBA) Banking Code of Practice. At present, the 'legislative and constitutional framework establishing these schemes do not allow first-e to obtain membership. Clearly, the 1987 UK Banking Act was framed long before the concept of Internet banking was envisaged as a means of offering consumers banking services across geographical boundaries of countries. Therefore, as the following comment shows, this may negatively affect consumer purchase intentions of first-e banking services, while all other banks are on a level playing field.

> *"First-E are not regulated by the Bank of England, as it doesn't have a UK banking licence. This means that in the event*

of a claim, you will be dealing with the Banque du France (start practicing your French!). Inevitably, the risks are higher."

(Steven Zuanella, iii).

This analysis has shown that security risk perceptions may have a significant influence on consumer purchase intentions.

8.5 Discussion

In many ways Internet banking is a process whereby banks delegate the tasks that they used to perform for customers. As the following testimony shows, customers perceive that as a result of this process, they are in control.

"I have been with Smile for nearly two months having moved from Abbey National. I find that Smile is first class. I had no problems moving my accounts and everything went very smoothly. I am now in full control of my banking from the comfort of my own chair. I think they have done an excellentjob and their web site is so easy and fast to use. Smiling broadly"

(David McAlister, iii)

On the other hand, banks are of the opinion that they are saving valuable resources. Both sides appear to gain from Internet banking. This perception has created an atmosphere where there is a very strong desire on the part of consumers to purchase Internet banking services, and banks on their part are willing to offer better terms to these customers compared to their terrestrial branch customers.

What is beginning to emerge in this exploratory analysis is that, price of the Internet banking service influences consumer purchase intentions to a very large extent. This variable coupled with a strong parent brand association and the time lapsed since the launch appear to be the three most important determinants of consumer purchase intentions. Additionally, the influence of service quality and security on consumer purchase intentions cannot be disregarded. The following discussion will examine each of these variables in turn, with a view of summarising the analysis and add more depth in substance where appropriate.

8.5.1 Pricing

The exploratory analysis has shown that the new entrants offered very attractive pricing regimes, at least initially. Moreover, in order to attract customers they gave assurances that their interest rates would remain above certain threshold levels for a designated period of time, thus may have dispelled some consumer risk perceptions. For example, egg, gave an assurance that from launch (October 1998) to January 2000 to maintain savings accounts interest rates at least 0.5 percent higher than the Bank of England base rate. Similarly, first-e also gave comparable guarantees. These interest rates were substantially above average market rates. The result was that customer demand for Internet banking services offering high interest was considerably higher than for services that offered much lower interest rates. This is one of the explanations of the very high customer demand for egg. However, towards late 2000 the interest rates offered by first-e and egg dropped in comparison to competitors. While the interest rate drops were by no means significant, they were no longer the best rates available on the market. There is evidence to suggest that a significant proportion of customers of both egg and first-e took their money elsewhere in search of higher interest rates. For instance, in September 2000, egg reported that the total customer deposit base declined by £800 million, from 6500 million (egg, 2000). While the reduction is less than fifteen percent, this re-confirms the earlier observation that price has a strong influence on consumer purchase intentions.

The above turn of events can also be argued as follows. The initial high interest rates and the subsequent drop in interest rates can be defined as a sophisticated form of price promotion. Winer's (1986) internal reference price theory postulates that the market share will decline after a promotion is withdrawn. The argument was that when price promotions are offered (in the form of above market average interest rates), consumers react positively by purchasing the high interest branded service product. Retraction of the deal, however, disappoints consumers and they may wait until the product (the account) is high interest bearing again to repurchase.

Therefore, it would be interesting to examine the extent to which these price promotions and subsequent withdrawals have had an influence on consumer purchase intentions. More specifically;

- What proportion of customers close their accounts and move them to banks offering higher interest rates?
- Related to above, are the reported reductions in deposit base due to a small number of customers moving large amounts of money, or a large number of customers moving relatively smaller amounts? As discussed in section (2.2.8), is customer inertia ('stickiness') so strong that only a minority of customers move their accounts? On the other hand, as argued below in section (8.5.2.2), would customer stickiness reduce in future?
- If a bank were to reduce interest rates then increases them again, do consumers go away and then come back?

These are some questions that require addressing in future research to gain a better understanding on the influence of price on consumer purchase intentions.

8.5.2 Brand and Risk

Brand is a commonly used extrinsic cue to infer and/or maintain quality perceptions. The importance of brand in banking services is more pronounced in that they are synonymous with trust. Further, it was concluded that brand is more institutional as opposed to being product specific. In the analysis, it was shown that there is some tentative evidence to conclude that brands in general influence consumer purchase intentions. More specifically, brand stability is signalled by the association with an established parent brand and time elapsed since launch of transactional Internet banking service.

8.5.2.1 Parent Brand Association

While it was concluded that there is some evidence to support that umbrella brands positively influence consumer purchase intentions, events during 2000 highlight some interesting developments. The creation of a 'new' brand for Internet banking service began in 1998. As Table (8.18) shows, this process commenced when Prudential, a well-known financial services brand with a banking license, launched egg. Having seen egg's success, Cooperative bank launched smile, in spite of already having an Internet banking service under the Cooperative Bank brand. The process intensified during 2000 with other well-established traditional banks launching new Internet banking services under different brands.

Name of Parent	Name of Internet banking service	Launch year
Prudential	egg	1998
Cooperative Bank	smile	1999
Abbey National Bank	Cahoot	2000
Halifax Bank	IF (Intelligent Finance)	2000

Table (8.18): Umbrella brands

While these subsidiaries acknowledge the parent brand, they also appear to be keen to establish a distinct brand on their own. A question arises, what is the objective of establishing a 'new' brand when the present brand is already well established? Several advantages of this strategy can be illustrated.

Firstly, parent banks are keen to avoid channel conflicts, the new brand can be used to mobilise 'new' customers that are cheap to service. As explained earlier in section (2.4.1), by serving customers only through the Internet, an Internet bank can make substantial savings.

Secondly, since these new customers are cheap to service they can be given preferential terms, compared to the parent bank customers. The new brand is important here since it helps to differentiate the customers, and consequently customers from the parent bank do not have grounds to claim that they are being given poor account terms.

Thirdly, while the link between parent and subsidiary is not very prominent, or not 'in-your-face', it is nevertheless given. Therefore the discerning consumer who is concerned about stability is informed and adequately assured. The perception is that these three reasons in combination will draw new customers and not a mere movement of existing parent bank customers into the newly branded Internet banking service, and will lead to bandwagon effects.

Fourthly, when competitors are launching new brand Internet banking services, there is a need to respond so that own customers are not lost to competitors. It is more prudent to cannibalise your own customers from parent to the new bank, rather than let competitors take them away altogether. Related to this, there is also the perception that city investors might perceive inaction to be a sign of weakness and lack of future strategy. This could drive share price down, and become the target of a takeover.

Finally, during late 1999 and much of 2000 there was a damaging price war among banks to attract customers with loss-leading account terms. Traditional banks, in their traditional state could not offer such accounts and suffer losses, but it appeared that new entrants could loose money during formative stages, and yet justify to investors and market analysts that such strategies are needed to gain market share. The downside, as noted earlier, is that the parent banks run the risk of cannibalising their own customer base, with their own customers becoming customers of the new brand. While some customers will inevitably become customers of both entities, the challenge lies in balancing the impact of all these factors, and arriving at a net gain both in terms of customer demand and financial profitability.

Taking a different perspective, it could be argued that it is inappropriate to analyse new Internet banking service entrants (those that are subsidiaries of traditional brands) under umbrella brands. As stated earlier, some of these new entrants do not prominently emphasise the link to the parent. Therefore it could be argued that they trying to create a different image, to cater for a certain segment of consumers. In chapter four it was stated that many people do not like banking. It was perceived to be an activity lacking friendliness, intimacy and conducted at conservative entities. The new brands are attempting to convey a sense of friendliness (at times using interesting if not peculiar names, for example the selection of 'egg' and 'smile', as the brand of a bank) and inject some fun and intimacy to such consumers. Crucially, as stated above, these new banks also offer very attractive account terms.

Then a question arises, if the aforementioned had been the strategies behind the creation of new brands, how successful will they be in the long run? The initial euphoria surrounding the launch and associated publicity of a new brand is bound to decrease in future. It is fair to state that the launch of egg created far more publicity than the launch of either Cahoot or IF. Equally, it is not possible to continue to offer loss leading account terms indefinitely, they will have to revert to more sustainable terms.

These are questions that cannot be answered in this exploratory analysis, but only addressed in the future with audited data. Such research may also address a notion highlighted in section (8.3.3), i.e. are the new Internet banking entrants 'trying harder' to meet customer expectations?

8.5.2.2 *Time Elapsed since Launch*

The significance of the time elapsed since launch of an Internet banking service in influencing consumer purchase intentions was discussed in the analysis. The following aspects also merit due consideration and thereby can add more weight to the conclusions in the exploratory analysis.

Throughout the analysis, it was seen that there is exceptional customer demand for egg Internet banking service. While some reasons for this demand were explained above in section (8.5.1), it may be important to reconsider them again. At the time egg launched in November 1998, it was the first and only new entrant offering Internet banking services. Admittedly, as explained in section (8.5.1), the interest rate offered by egg at that time was easily the highest rate of any savings account in the UK. Since it was the first among many, there was a considerable amount of excitement created which in tum led to increased customer demand. On the other hand, Barclays, BOS and RBS all offered very low interest rates. Yet, apart from egg they were the only banks able to attract in excess of 300,000 customers each. While accepting that traditional banks have an inherent advantage in persuading their customers to become Internet banking customers, another common denominator among these banks is that they were among the first to launch their transactional Internet banking services. Once a growing customer base is established, a bank may find that it becomes easier to influence prospective consumers to purchase Internet banking services from them.

Section (8.5.1) also highlighted that there has been an outflow from the deposit base of egg, with the drop in interest rates. The outflow in monetary terms has been considerable, signifying that prices are important consumers. However all the customers did not move, defying what some commentators who describe the purchase of Internet banking products from a competitor to be as simple as;

'a competitor is only a mouse click away'

(Hagel, et al., 1997. p. 47)

This customer reluctance to move out of banks may be attributable to the stickiness of banking customers discussed in section (2.2.8). This tentatively lends support to the notion that it is advantageous to commence operations earlier than competitors, since it influences consumer purchase decisions, and moreover, once a consumer becomes a customer there is a reduced likelihood of them leaving.

Additionally, the analysis of the content of the discussion forum contributions show that consumers consider the length of operations as an important consideration prior to purchase. Consumers tend to regard the time since launch as a surrogate for proven performance, reduced risk perceptions and stability of brand. Representative samples of such contributions appeared in section (8.3.5.3).

The date of reference for this research study was January 1, 2000 and many changes have taken since then. For instance, Internet penetration in the UK has increased to 32 percent of the population by the end of 2000 (Wright and Jayawardhena, 2001), and at the same time consumers have become more aware and used to dealing with Internet based services. Moreover the range of devices for accessing the Internet is constantly increasing. With the increased modes of access, faster and easier access, and last but not least with the increase in users customer stickiness may reduce. This might mean that the advantages that accrue presently to a bank due to time elapsed since launch, i.e. in terms of influencing positive consumer purchase intentions, may increasingly diminish in future.

8.5.3 Service Quality

The literature review illustrated the importance of service quality, highlighted some of the difficulties in defining and evaluating the construct, and also its influence on consumers. The debate on the interaction of service quality, consumer satisfaction and purchase intentions remains non-conclusive. Nevertheless, this research gave credence to the argument that service quality is an antecedent of consumer satisfaction, which in turn exerts a strong influence on purchase intentions. Thereafter it was demonstrated that in Internet banking services, service quality can be evaluated through the assessment of three characteristics of an Internet banking service.

1. Web site characteristics that include aspects such as navigation and design attributes that portray an image, akin to the physical facilities found in a terrestrial banking organisation. The analysis found that there is little evidence to conclude that there is a clear relationship between Web site characteristics and consumer purchase intentions.
2. The performance of a Web site and related components. Admittedly, performance is an all-embracing tenn encompassing the performance characteristics of the Web site, the downloading speeds of the Web site and the performance of the demo component. In this instance too, the analysis found that there little or no evidence to support the hypothesis that Web site performance may influence consumer purchase intentions.
3. The evaluation of responsiveness, quality and security features of communication. In the absence of face-to-face contact, aspects of speed, accuracy and responsiveness of communication through other means become of paramount importance. The importance and appropriateness of mystery shopping was pronounced in the evaluation of this aspect, and it was concluded that purchase intentions may be influenced by communication methods and related features.

8.5.4 Security Risk Perceptions

In this exploratory analysis it was tentatively concluded that there is some preliminary evidence to support the hypothesis that consumer purchase intentions are influenced by security systems incorporated into Internet banking services. Recent events may add some further support to this hypothesis.

During the period June - September 2000, there were a number of Internet security incidents that were given wide publicity in the media. Among them,

- a number of Powergen (a utility) customers who pay their bills using credit cards were able to observe credit card details of fellow customers,
- a number of Halifax share dealing customers have been able to access other customer details, and

- customers of Barclays bank could not log into their account for two days due a breach of security, this attracted considerable publicity.

A common denominator among all these incidents is that customers never lost money. However, they induced a new level of awareness of the possibility of security breaches in Internet related services. As a response many banks have revamped the presentation of security issues in their Web sites. In particular, banks have made their commitment to customers more lucid and also elaborated upon some of the technical aspects of enforcement of security. Perhaps it is fair to conclude that these events may have strengthened the relationship between perceived security risk and consumer purchase intentions.

Section (8.4.2) highlighted that first-e is operating in the UK with a French banking license and is based in Dublin, Ireland. As a consequence, first-e, is not currently a member of the UK Banking Ombudsman Scheme or a signatory to the British Bankers Association (BBA) Banking Code of Practice. The analysis of discussion forums uncovered that consumers attach a risk premium to this arrangement, and it negatively influences their purchase intentions. Additionally this bank has received a substantial amount of negative publicity from the media in general, and from the Consumer Association in particular. Perhaps in response to these concerns, in October 2000, first-e, set up a 'customer council' from within the organization. This is indeed a 'first' within the banking organizations. The task of this council is to investigate customer complaints and concerns. Admittedly, it may lack the perceived impartiality of the Banking Ombudsman Scheme. However it is evidence of the bank's admittance that at the present some consumers may not purchase Internet banking services from them, due to higher security risk in comparison to banks falling under the jurisdiction of the Banking Ombudsman. Therefore it further lends support to the hypothesis that consumer purchase intentions may be influenced by security risk perceptions.

8.6 Summary of Analysis

This exploratory analysis has shown that many hypothesised characteristics of Internet banking services influence consumer purchase intentions. Three sections of this chapter have been instrumental in drawing together these tentative conclusions. Firstly, section (8.3) examined characteristics associated with information search and acquisition stage of the purchase process. Secondly, section (8.4) illustrated characteristics related to experience qualities of the purchase process. Finally the discussion in section (8.5) added more substance to the findings of sections (8.3) and (8.4). It is prudent at this stage to present a summary of the influence of individual hypothesised variables on consumer purchase intentions. This is best presented by Table (8.19).

Hypothesis	Summary of tentative results
H 1: Purchase intentions, mediated by the perceived information search and experience quality, are influenced by the contents and presentation of the demonstration component of an Internet banking service.	Conflicting evidence. No obvious significant relationship.
H 2: Purchase intentions, mediated by the perceived information search quality, are influenced by the contents of an Internet banking service Web site.	A significant relationship is absent.
H 3: Purchase intentions, mediated by perceived technical qualities, are influenced by the functionality of banking service products of an Internet banking service.	No conclusion due to conflicting evidence.
H 4: Purchase intentions are negatively influenced by the prices (as measured by interest rates, fees, etc.) of an Internet banking service.	There appears to be a strong positive relationship, i.e. high interest rates for positive balances and low fees seem to influence purchase intentions.
H 5: Purchase intentions are influenced by the visibility (as measured by capture by popular search engines and linkages to other websites) of a brand name of an Internet banking service.	Little evidence of a relationship.
H 6: Purchase intentions are influenced by the association (or proximity) of an old established umbrella brand.	There is some evidence of a positive relationship.
H 7: Purchase intentions are influenced by the strength of a brand, as proxied by the time elapsed since brand launch.	There is evidence of a positive relationship.

H 8: Purchase intentions, mediated by the perceived Internet service quality, are influenced by Web-site characteristics.	There is little or no evidence of a significant relationship.
H 9: Purchase intentions, mediated by the perceived Internet service quality, are influenced by a system of components (as measured by the performance the Web site, communication methods, and related components).	There is little or no evidence of a positive relationship.
H 10: Purchase intentions, mediated by the perceived Internet service quality, are influenced by the responsiveness of communication, speed and accuracy of responses to customers.	There is some evidence a positive relationship.
H 11: Purchase intentions, mediated by the perceived Internet service quality, are influenced by Web site security.	There is evidence of significant positive influence.

Table (8.19): Hypotheses and summary of tentative results

8.7 Chapter Summary

The twin objectives of this exploratory analysis were: to illustrate the type of analysis that was feasible given the limitations in data, and to tentatively ascertain what characteristics many influence consumer purchase intentions.

Provisional eyeballing of individual right hand side (RHS) variables and their relationship with left hand side (LHS) variable, consumer demand, were tentatively reported, pending the availability of audited, time series estimates of the LHS variable. It appears that price of the Internet banking service influence consumer purchase intentions to a very large extent. This aspect coupled with branding, service quality and security appear to be four important determinants that influence consumer purchase intentions. Brand is characterised by the time elapsed since launch of the Internet banking service and the association with an established parent brand, i.e. umbrella brand. Service quality on the other hand envelops the characteristics of the Web site, performance of the Web site and the responsiveness of communication systems. It is important to reiterate that the findings of this exploratory analysis has been tentative and full empirical testing must await comprehensive and audited data. Nevertheless, this chapter laid the foundation on how future methodologies and studies in Internet banking and electronic commerce may evolve.

Chapter 9.

Conclusions and the Future

9.1 Introduction

Unlike with physical goods, consumer intentions to purchase services, and particularly financial services, have been shown to depend traditionally on intangible aspects of quality associated with face-to-face interactions between buyers and sellers. However, electronic commerce in general, and Internet retail banking in particular, ostensibly removes most human contact from service provision. This has profound implications for services marketing, and research methodology in particular.

In these novel circumstances, this thesis takes a first step toward understanding these implications for marketing research, by drawing attention to the methodological opportunities offered by Internet banking services. This study moves away from the traditional attitudinal research that has been used in previous research, and presents the development of the mystery-shopping research methodology as a novel and appropriate approach to understanding consumer behaviour in Internet banking services. It begins this process by formulating a research question that endeavours to identify the characteristics of an Internet banking service that influence consumer purchase intentions. It hypothesises that consumer demand for Internet banking is mainly a function of tangible service quality attributes, pricing and branding.

It is important to appreciate that this is the first rigorous academic study in this direction. An exploratory analysis on consumer purchase intentions and methodological implications are discussed in the relevant chapters. The objective of this concluding chapter is to summarise these conclusions and to propose directions for future research. Accordingly, the chapter is divided into five sections. The first and second sections summarises the main aspects of the literature review, i.e. the characteristics of the banking sector and the consumer purchase process. The third and perhaps the most important section, presents the development, appropriateness, academic rigour of the mystery-shopping methodology in a concise manner. The fourth section highlights the contribution of this thesis: by re-emphasising the importance of the development of the mystery-shopping methodology and presenting a summary of the exploratory analysis of the characteristics of an Internet banking service that would influence consumer purchase intentions. The fifth and final section presents proposals for the future development of the mystery-shopping research methodology and policy suggestions for Internet banking services.

9.2 Characteristics of the Banking Sector

Many of the traditional underlying reasons for financial intermediation are being eroded by changes in the external environment and are further compounded by changes taking place internally within the banking industry. While these changes in the banking sector have adversely affected traditional banks, they have paved the way for the emergence of new entrants.

The challenge that lies ahead for banks is fourfold. Firstly, they need to satisfy customer needs that are complex and difficult to manage. Secondly, they need to face up to increased competition from within the sector and from new entrants coming into the banking sector from disparate sectors, for example, retailing. Thirdly, they need to address the demands placed upon the supply chain. Finally, they must continually invent new products and services in light of the changes described above. Central to these challenges are delivery strategies for banking services.

Internet banking is presented as both a viable alternative and complimentary distribution channel with the potential to resolve many of the inherent disadvantages of traditional branch banking. It is more than just another delivery channel for existing products and existing customers. The Internet provides many opportunities for banks. The most significant advantages of the Internet are cost savings, opportunities for increasing the customer base and an ability to provide the service free from temporal and geographical limitations. A related aspect is that an Internet bank can act as a facilitator in Internet payment systems or as a provider of other services, thus assisting the growth of electronic commerce in general (Jayawardhena and Foley, 1999; Jayawardhena 1999a). A Web site is a powerful marketing tool with comparatively low outlays and can be used to promote other products and services to generate advertising revenue. It provides an ideal opportunity to further strengthen the terrestrial bank's brand as well as developing an online brand.

9.3 The Purchase Process

9.3.1 Buyer Behaviour Models

Most models of buyer behaviour provide normative rather than strictly quantitative explanations of buyer behaviour. Many buyer-behaviour models were discussed, but there is no single explanation of consumer behaviour and intuitively elements from all these approaches can be recognised in one's own behaviour. Hence, the model revolving around the concept of the cognitive consumer, initially developed by Howard and Sheth (1969), is adopted as the framework for analysis. This model suggests that the purchase process is a sequence of tasks characterised as a problem solving exercise. It is a generic model that can be applied to both goods and services, and provides a rich information-processing perspective.

The first stage is the increasing recognition of a need or want that can be fulfilled through some form of consumption. The second stage commences with an information search phase, followed by an evaluation stage where consumer engages in some sort of an evaluation process to arrive at a ranking of the identified product alternatives. The unique characteristics of services (for example, intangibility, etc.), emphasise the role of information in this pre-purchase stage of services. While these two stages occur traditionally prior to the contact with the service provider, the third stage is represented by the purchase, where the consumer comes into contact with the service provider. The fourth and final stage is the post-purchase evaluation, i.e. the evaluation of service quality. In the case of service products, unlike in the case of physical products, the latter two stages occur simultaneously, for services are consumed while being purchased, hence the evaluation of service quality take place during the purchase process.

This thesis identifies the critical roles of service quality and satisfaction in the formation of consumers purchase intentions. The debate on the exact nature of how the relationship between service quality and satisfaction combine to impact on consumer purchase intentions remains unresolved. This is because empirical efforts to validate the specific nature of the relationship have supported both possible relationships between constructs, i.e. that service quality leads to satisfaction that in turn influence purchase intentions, and/or that satisfaction leads to better perceptions of service quality, which in turn influence purchase intentions. After a critical examination of both arguments, the stance taken in this thesis is that service quality is an antecedent of consumer satisfaction, which in tum exert a strong influence on purchase intentions.

9.3.2 Variables influencing Purchase Intentions

Having identified a research framework that encapsulates the purchase process, it is necessary to recognise variables that influence purchase decisions. Empirical research work in financial services identifies the influence of a number of variables on consumer purchases. While price is perceived to have an important influence on consumer purchases, there are very few studies that examine this aspect. The most notable contribution comes from Leonard and Spencer (1991), they discuss the importance of the pricing of financial services, and conclude that it is important to convey the perception of value in a bank's prices. They propose that such an objective can be achieved by creating the 'right' pricing environment, through research and pricing committees.

Some authors argue that when buyers cannot easily evaluate the qualities and value of the service or capabilities of the service provider, then brand reputation may serve as an important proxy for more detailed evaluations.

A large number of studies examine the relationship between purchase intentions for financial services and service quality, and the measurement of service quality in financial services. They emphasise the roles played by both technical and functional dimensions of the service. A number of these studies also find that the performance of the delivery of a financial service has an impact on service quality perceptions.

The crucial question is whether these traditional variables influence consumer purchase intentions in Internet banking services to the same extent, and if so how can these be measured? This necessitated an assessment of Internet banking service characteristics. This examination reveals that Internet banking services dilute some of the traditional qualities of service products, and show qualities that are more analogous to those of physical products. Furthermore, additional qualities that are integral to traditional financial services, *i.e. fiduciary responsibility, two-way information flows* and *long term relationships* are somewhat blurred in Internet banking services. It is therefore hypothesised that consumer purchase intentions in Internet banking services are influenced by information consisting of both presentation and content, technical and functional qualities, prices, branding, service quality and security of an Internet banking service.

9.4 The Research Methodology

How can these hypotheses be tested? In its attempts to address this issue, this thesis takes a first step towards understanding the implications of the Internet for marketing research and to propose a new methodology. It moves away from the traditional attitudinal research that has been used in previous research, and suggests mystery-shopping research methodology as a novel and more appropriate alternative. It is important to appreciate that this is the first rigorous academic study in this direction, and while a considerable amount of attention is paid to the development the research instrument, there are opportunities for further refinement. The findings of the study are presented as an exploratory analysis at this stage. These may be refined in many ways in future with the availability of audited data and better understanding of the process carrying out commercial activities on the Internet, and in particular the delivery of services.

9.4.1 The Background

This discussion is necessary in order to highlight many shortcomings in adopting existing methodologies to the Internet. Pervious researchers utilised a combination of questionnaire surveys, interviews and case studies to examine research questions dealing with traditional financial services. There may be an argument to adopt these data collection methodologies to the Internet with appropriate modifications.

Such arguments may be made in light of many potential advantages of the Internet for research, in particular for data collection methodologies. For instance Johnston (1999) concludes that these advantages include greater speed, lower cost, improved accuracy in encoding data and the ease of obtaining international samples. However, Pincott and Branthwaite (2000) assert that these claims are seldom backed up by detailed, comparative analysis of all the costs (preparation of data collection tools, collecting data, analysis, etc.) and caution should be exercised in assuming that the benefits are universal and apply to all forms of research. Others such as Kent and Lee (1999) and Schillewaert, *et al.*, (1998) subscribe to this view by contending that, in emphasizing the marketing opportunities and potential of the Internet, reviews tend to gloss over the problems and limitations, which include poor response rates and umepresentative samples (that lead to inaccuracies in assessing consumer attitudes and reactions).

Internet respondents can be more self-centred and self-absorbed. Such traits can be positive and Mehta and Sivada (1995) assert that electronic responses to open ended questions are up to three times longer, contain clarifications and explanations, and have more candour and are less inhibited. Scholl, *et al.*, (1999) present findings that are at the opposite end of this spectrum, and thereby raising concerns to counter the arguments in favour of using the Internet for data collection. Research by Adriaenssens and Cadman (1999) finds that informal, virtual contact between respondents and moderator can result in respondents changing their demeanour, and, by extension, this could affect the quality of responses. Related to this, there is widespread disagreement on the effects of moving from interviewer-administered, face-to-face questionnaires to self-completion questionnaires with no research personnel to help and advise (Strauss and Frost, 1999). Consequently, results need to be treated with caution.

An Internet bank attaches great value to the captive consumer base, as a consequence these Internet banks will not usually give researchers email or any other form of access to its consumers for the purpose of carrying out attitudinal research. This is because the findings of these attitudinal surveys can be highly commercially sensitive.

Even with direct access to consumers there are several threats to the generalizability and replicability of surveys on Internet based services. Firstly, when respondents respond to a questionnaire, the respondent's PC including hardware specification, configuration and installed software, quality of the connection to the Internet, etc. will have a major influence on their responses. It is very difficult, if not economically impossible to maintain the required uniformity for the results to be generalised. Secondly, data gathered directly from consumer Web-site observations must appreciate the ephemeral nature of Web sites. Consequently, attitudinal surveys using consumers at two different times will not result in replicable outcomes, and presents a strong case against carrying out attitudinal surveys on consumer perception on Internet banking services delivered over the Web. Thirdly, in a typical household the person who buys banking products may not be the only person who consumes it and similarly the decision of preciously what to buy may be taken in consultation with other members of the household. As a result a simple questionnaire type of instrument may not capture all influences concerning a purchase. These shortcomings would prevail irrespective of whether the attitudinal data collection method is Internet based (i.e. a Web questionnaire, email questionnaire, etc.) or traditional (i.e. a postal questionnaire).

Additional problems persist for Internet administered data collection methodologies. Most individuals assume that they are anonymous on the Internet, and they regard their privacy as the most important element (Strauss and Frost, 1999). There can be several consequences of this: (1) there is a tendency for individuals to adopt extreme positions, (2) this apparent lack of identity can also result in an individual submitting multiple responses (Pincott and Branthwaite, 2000), (3) refusal to provide information. For instance, according to Hoffman, *et al.*, (1997), 94 percent of Web users have refused to provide information at a Web site. Perhaps one of the most significant findings of research by Hoffman, *et al.*, (1997) is that 40 percent of users give false information.

While this commentary highlights some of the challenges faced in carrying out research on the Internet, all is not lost since researchers have unparalleled access through Web sites to service characteristics previously unavailable. There may be some potential advantages that may be of considerable benefit.

Internet services offer enormous new opportunities for carrying out research. Many Internet services, as shown in the case of Internet banking, involve the routinization of several aspects of service provision, this reduces the difficulties involved with the measurement of intangible qualities. Lowe and Kuusisto (1999) assert that the increasing tendency of retail banks to deliver their services via information technology-driven channels, easily leads to the commodification (Pine and Gilmore, 1998) of their previously high value added services. Furthermore, although direct surveys of consumers are virtually ruled out, it is much easier for researchers to simulate the consumer's search for and perception of an Internet banking service by simply logging on to a Web site and making observations. In effect the conditions are ideal for the revival of mystery shopping methodology on a wide scale on Internet based services.

At this early stage in the development of Internet banking, it is only possible to perform the first step in this commercial and methodological environment, i.e. to propose a methodology and suggest first, tentative results.

9.4.2 Mystery-Shopping

The research methodology developed in this thesis is a derivation of participative observation in the terrestrial world. Although the use of mystery-shopping in the academic world has been limited, it has been extensively used in terrestrial banking in a commercial context.

Internet banking services involve the routinization of many aspects of service provision, which to some extent eliminates the difficulties involved with the measurement of intangible qualities. Hence, as explained above, in comparison to direct attitudinal research surveys of consumers, it is much easier for researchers to simulate the consumer's search for and perception of an Internet banking service by simply logging on to a Web site and making observations. In essence the mystery-shopping methodology is a research instrument comprising of a series of Internet scorecards which makes it possible to carry out detailed cross-sectional and time-series evaluations of Internet banking service Web sites and associated features.

9.4.2.1 Advantages of Mystery-Shopping

Advantages of mystery-shopping in Internet banking services can be summarised as follows.

1. In contrast to attitudinal surveys of customers, the mystery-shopping approach is being used to measure the process rather than the outcomes of a service encounter (Wilson, 1998). The emphasis is on the service as it unfolds, looking at which activities and procedures do or do not happen rather than gathering opinion about the service experience (Miles, 1993).
2. The literature review in section (4.3) revealed that consumers of banking services are more interested in functional qualities, as opposed to technical qualities. Hence mystery-shopping is a very suitable methodology (for an innovative application such as Internet banking services) as it provides more objective measures of service delivery.
3. Desk-based, mystery-shopping of the Internet banking service Web sites and demos produces very detailed measures at a fraction of the cost of other methods, such as an attitudinal customer survey. The significant variable costs are limited to the wage costs of a researcher and the Internet connection costs. Incidentally, Internet connection costs have continued to fall significantly during the latter half of 2000, and the indications are that they will continue to fall for the foreseeable future.
4. The data gathered in this method are largely superior to data gathered through other forms (traditional) of mystery-shopping, since measurements rely less on the subjectivity and competence of the researcher.
5. Related to above, while a significant number of measures may be still be subjective in nature, many freely available software tools on the Internet (for example, Alexa monitor) can be used to enhance measurements, and by extension enhance the objectivity of measures.
6. Unlike other methods of data collection, such as a customer attitudinal survey, the data collection process and the data entry process can take place simultaneously. Such an arrangement saves a considerable amount of time, money, and moreover it would mitigate tendencies for any data transfer errors. Thus, this course of action speeds up the entire research process substantially.
7. Unlike field research, desk-based research affords more opportu-

nities for controlling the effects of the external environment. Additionally, such influences from the environment, if they do occur, can be monitored and recorded. Where controls are impossible, for example variable login times, time spent on reading the information on Web pages, etc., variations can be monitored and fed into the database, so that allowances can be made at the analysis stage.

8. In a consumer survey, the respondent's PC including hardware specification, configuration and installed software, quality of the connection to the Internet, etc. will have a major influence on their responses, hence raising the importance of maintaining uniformity. It is very difficult, if not practically impossible, to maintain the required uniformity among a representative sample of respondents for the results to be generalised. This methodology avoids this aspect altogether because only one PC is used for the entire process.
9. Internet banking is a dynamic sector, where processes, procedures, technology, products, etc. all change much faster than in the terrestrial banking sector. Hence a research methodology employed to measure an aspect of Internet banking service must have the inherent speed and flexibility at its core to be a credible and worthwhile methodology. The qualities of Web site mystery-shopping qualify as a more suitable methodology on these grounds.
10. Related to the above, in commercial applications, where there is a need to conduct frequent assessment of competitors, Internet mystery-shopping can be used very effectively and adhere to MRS code. Internet mystery-shopping avoids the MRS code limitations imposed on frequent assessments of competitors in the terrestrial world. These include: ensuring a spread of mystery-shopping assessments over time; assignments not all in one week/one day; not carrying out mystery-shopping exercises at busy periods, practical limits on the frequency of mystery-shopping visits to individual outlets, etc. (Miles, 1993). All these limitations are avoided by Web site mystery-shopping.
11. It offers total anonymity, and is a true 'mystery' shopping scenario with the service provider not being aware that they are being shopped and therefore unable to make 'special' arrangements for the 'mystery' visit.
12. It also avoids some of the ethically contestable considerations of a traditional terrestrial mystery-shopping exercise. The resources of a bank are insignificantly wasted on Web site mystery-shopping. In practical terms the mystery-shopping evaluation does add to the

number of users using the servers of the bank, thereby increasing traffic. This is because, the marginal increase in traffic is negligible when most of these back end servers are typically designed to handle more than 5,000 users at a given time.

Perhaps the strongest argument in favour of mystery-shopping in Internet banking services is that subsequent banking experience is likely to be identical to a trial experienced though the browsing process of the Web site and observing a demonstration. In most traditional services a trial is not a common option. Although it is possible to try one service episode to establish whether one will purchase a full service or a sequence of episodes, there is no guarantee that future service experiences will be identical to the trial.

9.4.2.2 *Weaknesses of Mystery-Shopping*

Weaknesses of mystery-shopping in Internet banking services can be summarised as follows.

1. The entire outcome of the exercise depends on the quality of the data collection instrument. Careful design is required to give comprehensive coverage. The omission of a crucial factor may yield spurious results.
2. Because the methodology does not solicit views from a sample of respondents, it lacks the diversity of open-ended answers and suggestions.
3. Inevitably, some measures in a mystery-shopping exercise produce subjective evaluations. While it is impossible to remove the subjective element, it can be reduced by formulating appropriate measures that have objective underpinnings.

On balance, taking into account all factors, it is fair to state that mystery-shopping provides new opportunities in a number of areas for research.

9.4.2.3 *Mystery-Shopping and It's Methodological Rigour*

Methodological justification requires an assessment of how the proposed methodology measures up to indicative criteria governing good research design in respect of measurement, validity, reliability, generalizability and relevance of the measuring instrument.

Measurement

Taking into account the particular problems associated with ordinal scales in previous studies, a graphic positioning scale is used to assess the variables. A bi-polar semantic scale is incorporated. The scale, 1Ocm long, is calibrated at one-centimetre intervals, and provides a point of reference for the researcher from which to evaluate the subjective assessment of variables. Such an ordinal level of measurement is important for it retains its uniqueness up to a monotonic transformation, i.e. any order preserving transformation does not change the information contained. A considerable amount of attention is paid to minimise measurement errors, and the design of the bi-polar semantic scale lends to this process, and these are discussed below under validity and reliability. A slight drawback of the scale is that data entry takes longer, for the exact location of the evaluations must be carefully read of£

Validity

As stated above, validity is concerned with techniques for reducing measurement errors. Three types of validity are distinguished, content validity, empirical validity and construct validity, each of which considers a different aspect of measurement.

There are two common varieties of content validity: face validity and sampling validity. Mystery shopping scale measures were found to satisfy face validity. This is because it has (1) provision for replicability, (2) is easy to understand and (3) scale measures are developed after an extensive literature review. It also satisfies sampling validity, in view of the fact that the entire population, comprising of fifteen operational Internet banking services in the UK as the date of reference (January 1, 2000), are analysed in this research.

Empirical validity, the relations between a measuring instrument and the measured outcome, of the mystery-shopping methodology is tentatively assessed in the exploratory analysis. However, due to limitations of the data collected the opportunity for scrutinising the measurement instrument against some of the desired outcomes is limited.

A construct can be validated by observing whether predictions made on the basis of those propositions about its relationship to other variables are confirmed when tested. Within the exploratory analysis, some of the sources of evidence yielded similar results, and therefore it can be concluded that there is convergent validity in the construct. In this instance too, data limitations hampered the rigour with which such testing could take place.

Reliability and Generalizability

Reliability is very important to any research, and this research is no exception. This research design can be criticised that it lacks reliability, since there are no opportunities to observe whether there are any variable errors that differ from one observation to another. This is a valid criticism at this present moment, and the potential for this relative lack of reliability can be considerably improved with the availability of audited time series data. Steps are taken to ensure the generalizability of this study. It is possible to argue that this study is based on simulation, and therefore lacks practical credibility of findings, but simulation in the case of Internet banking services equates to actual service delivery.

9.4.2.4 Implementation

The mystery-shopping instrument comprises of a series of interrelated scorecards, evaluating an Internet banking service on four levels, (1) Web site evaluation, (2) demonstration component evaluation, (3) email/query application evaluation and (4) overall evaluation. These scorecards in combination evaluate characteristics of an Internet banking service that influence purchase intention by examining the sequence of tasks a cognitive consumer goes through in making a purchase. This sequence was summarised in section (9.3.1).

In developing these Internet scorecards, which evaluate variables generated to test the hypothesis, several aids have been used. Firstly, like any academic research, a comprehensive literature review, summarised very briefly in the first two sections of this chapter, formed the foundation of the research instrument. Secondly, Internet discussion forums are used to both assist in the generation and refinement of measures and to add credibility to the research findings. Thirdly, online surveys and consultancy reports are used for similar reasons. The validity of the use of discussion forums, online surveys and consultancy reports in this research study was dealt with in detail. This discussion concludes that although there may be deficiencies in these aids, the benefits that they add to the study outweigh any disadvantages that are inherent in them.

As stated above, due to the ephemeral nature of Web pages a copy of all Web pages were stored in a Compact Disk Recordable (CD-R) for analysis and replicability. Internet Explorer (IE) version 5.0 is used for browsing and downloading since this Web browser met all critical requirements, in particular with regard to downloading of images. Subsequently, each of the bank Web sites was evaluated using the mystery-shopping scorecards on four levels as stated above. During this evaluation, every effort was made to simulate a consumer seeking to purchase an Internet banking service.

9.4.3 Methodological Implications

Internet banking involves a number of unique features that generate significant problems and opportunities for marketing and other researchers. On the face of it, the novelty of Internet banking makes time-series observations infeasible, until the industry matures. On the other hand, researchers who obtain data on RHS service characteristics directly from Web site observations must recognise the ephemeral nature of Web sites and the fact that they are modified frequently, usually without leaving a trace of preceding offerings. Researchers therefore have a duty to store observations to aid subsequent replicability.

In addition, the fact that customer bases are themselves a massive asset to Internet banks, and a significant proportion of their total market value, itself has two negative implications for researchers. First, as discussed above, it is unlikely that even one Internet bank will give academic researchers email access to its customers for purpose of attitudinal surveys, though of course in-house surveys are common [Dammeyer and Moriarty, 1999]. In the UK, statutory data protection laws make this process even more difficult. Second, estimates of the number of customers enrolled will be important and confidential information for banks, and significant 'bandwagon' effects (Moschandreas, 1994) provide a motive for banks to over-estimate the true size of their base of active customers, for example, through the continued inclusion of inactive customers. This makes triangulation (Yin, 1993,) particularly desirable in the context of auditing banks' own sales estimates.

At the same time, Internet services offer enormous new opportunities for market researchers. Such services involve the routinization of many aspects of service provision, and this reduces the difficulties of measuring previously intangible service qualities. Furthermore, although direct (other than in-house) surveys of customers are virtually ruled out, it is much easier for researchers to simulate the customer's search for and perception of an Internet service by simply logging on to a Web site and making observations.

9.5 The Contribution of this Research

This research has made a very important contribution to the advancement of knowledge. It has developed and presented the mystery-shopping methodology as a novel and appropriate approach to examine consumer purchase intentions in Internet banking services. The preceding section of this chapter summarised the development, appropriateness and the level of academic rigour that is inherent in the mystery-shopping methodology.

As acknowledged and frequently stated on several occasions times throughout this thesis, the methodology proposed represents the first steps towards a new methodology in marketing research. Any research methodology of a piece of research encompasses the research instruments (the mystery-shopper Internet scorecards), the data collection process, the data analysis and the reporting of the findings. While research instrument and the data collection process are sufficiently academically rigorous, the data analysis stage in particular suffers from a number of problems which need to be resolved. The two most important shortcomings are the lack of audited data and lack of time series data. It is inevitable that therefore the findings of the research are equally handicapped, and therefore this research only presented an exploratory analysis of the characteristics of an Internet banking service that influence consumer purchase intentions.

The next section examines how the research methodology as a whole can be improved.

9.5.1 How can the Methodology be Improved?

Above all, audited customer demand information is required. Even then however, the methodology may still not result in robust conclusions. This is because, an institution needs time to establish itself, and this is especially so with a banking institution. Out of the new entrants, smile and first-e and, from the 'traditional' banks, BOS, NatWest, Halifax and First Direct all started Internet banking services after September 1999. The reference date of this study, January 1, 2000 is too close to service launches to reasonably estimate the determinants of demand. It follows from this that the methodology can be significantly improved ultimately by examining time series data.

In the long term, a methodology for analysing the demand for Internet banking services would involve five stages.

1. Researchers' surveys of Web site characteristics.
2. The formation of representative panels of customers to develop measures and weightings for the measurement of Web site characteristics and supplier responsiveness to queries.
3. Factor analysis to combine clusters of characteristics
4. The collection of audited data on actual banking sales from Web sites.
5. Full time-series regression analysis to identify significant influences on actual purchase decisions.

Having examined the methodological contribution in detail, it is important to summarise the main findings of the exploratory analysis, described above.

9.5.2 Characteristics of Internet Banking Services that Influence Consumer Purchase Intentions

An exploratory analysis of the hypotheses show that consumer purchase intentions in Internet banking services are influenced by pricing, time elapsed since launch, association with a stable parent brand, service quality and security. It is important to appreciate that these conclusions are tentative.

Price

It was established that Internet banking service prices has the most influence on purchase intentions. The analysis drew much support to this conclusion from the customer demand for the low priced (high interest) products offered by the new Internet banking entrants. Subsequent increases in prices (lowering of interest rates), and the reduction in the customer base added strength to this conclusion. These events, i.e. low prices initially and then increasing them, can be interpreted as a sophisticated price promotion. Therefore it was suggested that if price is indeed a characteristic of an Internet banking service that would strongly influence consumer purchase intentions, it might be possible to explain consumer purchase intentions through Winer's (1986) internal reference price theory. This theory postulates that the market share would decline after a promotion is withdrawn. If this argument were to be proven in future (once time series audited data is available), it would add strength to these tentative conclusions.

A further observation that may be applicable is that, as discussed below, as customer stickiness declines the importance of price in influencing consumer purchase intentions may increase.

Time Elapsed since Launch of Internet Banking Service

The significance of the time elapsed since the launch of an Internet banking service in influencing consumer purchase intentions was discussed at length. It was observed that the excitement at the launch of a novel service may influence consumer purchase intentions to a significant extent. It was also seen that once a growing customer base is established, a bank may find that it becomes easier to influence prospective consumers to purchase Internet banking services from them, due to bandwagon effects. Support for this conclusion was also found in the analysis of forum contributors', which showed that consumers tend to regard the time since launch as a surrogate for stability and trust. There was also evidence to show that consumers tend to regard the time since launch as a surrogate for proven performance, reduced risk perceptions and stability of brand, leading to positive purchase intentions.

The previous section argued that consumer inertia may be high even in Internet banking services, and therefore it is a considerable advantage for a bank to commence Internet banking service at the earliest possible instance. However, with the range of devices and speeds for accessing the Internet constantly increasing, coupled with the number of Internet users increasing, customer stickiness may reduce. Currently the time elapsed since launch does appear to influence consumer purchase intentions. However, its influence in the future might decrease.

Parent Brand Association

There is some evidence to support the assertion that umbrella brands positively influence consumer purchase intentions. This discussion prompted a question about why there was a perceived need by banks to create a subsidiary with an unknown brand when the parents were well respected financial institutions in the UK. Only two banks in this research, first-e and BankNet, lacked the parentage of a well known UK financial institution. There were found to be a number of reasons behind the launch of 'new' brand Internet banking services.

Firstly, it was argued that they are trying to create a different image, to cater for a certain segment of consumers. Traditional banking is perceived to be an activity lacking friendliness, intimacy and conducted at 'conservative' entities. The new brands are attempting to convey a sense of friendliness. Secondly, parent banks are keen to avoid channel conflicts, the new brand can be used to mobilise 'new' customers that are less costly (compared to branch customers) to service. Thirdly, since these new customers are cheap to service they can be given preferential terms, compared to the parent bank customers. Fourthly, while the link between parent and subsidiary is not very prominent, but discreetly represented. Therefore the discerning consumer who is concerned about stability is informed and adequately assured. Finally, during late 1999 and much of 2000 there was a damaging price war among banks to attract customers with loss-leading account terms. Traditional banks, in their traditional state could not offer such accounts and suffer losses, but it appeared that new entrants could loose money during formative stages, and yet justify to investors that such strategies are needed to gain market share. This was possible because of the particular market sentiment during that period. During much of 1999 there were wildly optimistic predictions for any Internet based businesses. One of the most significant characteristics of any business during this period was the number of customers that a business was able to attract, i.e.

the larger the customer base, greater the perceived 'worth' of the business. The success or failure of an Internet business was measured against this criterion by the media and investors, and Internet banks were no exception to this rubric

These reasons in combination appear to have influenced consumer purchase intentions. However, because of the very factors giving rise to new brands are not sustainable indefinitely in their present form, concerns were raised about how successful they will be in the long run. Especially once the initial euphoria dies away and once the loss leading account terms revert to more sustainable terms. This is a question that can only be answered in the future with audited data.

Service Quality

The importance of service quality, difficulties in defining and evaluating the construct, and also its influence on consumers has been discussed at length in this research. Within the wider academic community, debate on the interaction of service quality, consumer satisfaction and purchase intentions remains non-conclusive. This research gave credence to the argument that service quality is an antecedent of consumer satisfaction, which in tum exerts a strong influence on purchase intentions. The influence of service quality in influencing consumer purchase intentions was evaluated through the assessment of three levels of characteristics of an Internet banking service.

Firstly, it was found that there is little evidence to conclude that there is a clear relationship between Web site characteristics and consumer purchase intentions. Web site characteristics include aspects such as navigation and design attributes. These were argued to portray an image akin to the physical facilities found in a terrestrial banking organisation, and therefore the above conclusion was unexpected. Secondly, it was found that there is tentative evidence to support the hypothesis that Web site performance may influence consumer purchase intentions. Performance, an all-embracing term, encompassing operational characteristics of the Web site, downloading speeds of the Web site and the performance of the demo component. And finally, it was tentatively concluded that purchase intentions are influenced by communication methods and related features. In the absence of face-to-face contact, aspects of speed, accuracy and responsiveness of communication through Internet means is of paramount importance. The importance and appropriateness of mystery-shopping was pronounced in the evaluation of this aspect.

Security

There is some evidence to support the hypothesis that consumer purchase intentions are influenced by security systems incorporated into Internet banking services. High profile Internet security breaches during the year 2000 may have induced a new level of awareness of the possibility of security breaches in Internet related services. As a response many banks have revamped the presentation of security issues in their Web sites. In particular, the banks have made their commitment to customers more lucid and also elaborated upon some of the technical aspects of how security is enforced.

The setting up of a 'customer council' within first-e is also of significance. It was argued that this is evidence of the bank's admittance that at present some consumers may not purchase Internet banking services from them, due to higher security risk in comparison to banks falling under the jurisdiction of the Banking Ombudsman.

Perhaps it is fair to conclude that these events have strengthened the relationship between perceived security risk and consumer purchase intentions.

9.6 The Future

Internet banking services are in their infancy. Banks, consumers, policy makers and researchers are all on a learning curve. The coming years will make the strategies of banks more clearer, and consumers will become more discerning in their purchases of Internet banking services. These in combination will define the true extent of the impact of Internet banking in the financial services sector. Policy makers have the responsibility to foster and promote an environment that protects the interests of all parties. Researchers have an opportunity to improve existing research methodologies and contribute to the understanding of this important service. While the ways in which researchers could improve research methodologies has already been discussed, the following section examines how policy makers could achieve these objectives.

9.6.1 Policy Suggestions for the Internet Banking Services Sector

The banking services sector is a very important sector of the economy. It is of paramount importance that such a significant sector operates within a framework that (1) ensures rights and obligations of all parties concerned, (2) promotes the sustainability and ensures the growth of the sector. At the present Internet banking services comprise only a small proportion of the banking sector, and consequently it operates within the terrestrial framework. With the growth of Internet banking, the inadequacies of the framework in its present form may surface. This is because, as highlighted on numerous occasions throughout this thesis, some of the qualities of Internet banking services set them apart from traditional banking services. The following are some suggestions that may be applied to refine existing frameworks.

Regulation

Internet banking is currently regulated under the regulations that apply to terrestrial banking services. It has been shown on occasions that there may be inadequacies with such an arrangement. Hence it may be prudent to introduce new measures or change current measures along the following suggestions, which recognise the unique nature of Internet banking services.

1. Regulatory bodies may have to consider and take action to provide audited estimates of customer demand, etc. Besides aiding research, customers are entitled to this type of information, since they may need to be satisfied that a particular Internet banking service is credible.
2. Current regulations require that all customer identities be physically verified (for example, against a statement from a utility provider or other appropriate document). They may be modified to accommodate electronic verification. It is possible to carry out this verification process with, for example, by collaborating with a credit-referencing agency. Physical certification may be only employed only where electronic means fail.
3. In general regulations may be made flexible to reflect new methods of working and take into effect the speed of implementation, with adequate safeguards. For instance, with '24x7' operations, a 'working day' may have to be re-defined.
4. Banks like first-e, which operate in the UK with a banking license from another European Union country, should be encouraged to join the BBA and come under the jurisdiction of the Banking Ombudsman.
5. Related to above, banks may be encouraged to offer their services across geographical boundaries. Therefore, any bank offering services in a particular country may be required to join the appropriate banking regulatory body of that country. This process may increase competition in the banking services sector across countries and bring down the cost of financial intermediation in general.

Changes in Procedures

Although Internet banking represents a major advancement in banking procedures, it appears that they are operating under procedures that apply to terrestrial banking services. Clearly changes are required. A change in procedures in one banking services over another (i.e. terrestrial services and Internet services) is bound to create problems and is undesirable. Hence, it may be prudent to limit changes along the following suggesting, these avoid conflict and benefits both terrestrial and Internet banking customers.

6. The 'clearing' (for the effects to be realised) of cheques may be made quicker. At present it takes five working days for a cheque to clear. This is based on technologies and methods of working that existed more than 20 years ago. As a result of these delays, banks earn interest at the expense of customers. Such a change would benefit both Internet and terrestrial banking service customers.
7. To take full advantage of the capabilities of the Internet, real time accounting may be made available to customers. For example if a customer makes a transfer from one account to another, in most cases, it takes a day to reward the transaction. With real time accounting, as the name implies, customers are rewarded with the transaction instantaneously. To ensure that terrestrial customers are not penalised, their transactions too could be processed at the time of making the transaction.

Security and Access

The significance of security has already been illustrated. Additional measures suggested here are important.

8. Web sites and interfaces may be made accessible to all potential customers. While recognising that there is a need to use the latest technologies for security purposes, a concerted effort must be made to use the simplest forms of technologies (for example devoid of additional plug in software), conform to accepted guidelines in authoring, without compromising security. This may be of paramount importance when the use of non computer based devices (for example, Internet enabled mobile telephones) to access the Internet increases.
9. Security measures may have to be increased. This has been necessitated due to wide publicity given to incidents in the latter half of 2000. Several examples have been highlighted. All these problems share a common thread, in that customers' security was compro-

mised. Although no one suffered any financial loss, security systems in place should be made more robust, and efforts made to convey these added measures to customers. Security is one of the most important, if not the most important perceived aspect of Internet banking services. Consumers need be to be assured that Internet banking services is safe. This process of confidence building can only take place if there are no high profile 'security incidents' reported in the media. Even minor incidents are reported disproportionately.

10. Some banks, for example Nationwide, require a customer to remember up to three security 'passwords', comprising a series of numbers determined by Nationwide themselves. While recognising the need for passwords for security purposes, it makes life easier for customers to choose their own passwords. If a majority of banks can accommodate this option, there should not be any insurmountable technical difficulties. It is also important to appreciate that this recommendation should in no way compromise the security.

Internet banking services is both an important and challenging area for researchers. This thesis has developed and presented the mystery-shopping research methodology as a method of examining the characteristics of Internet banking services that influence consumer purchase intentions. While making a contribution to the understanding of consumer purchase intentions, this thesis has laid the foundation for the enhancement of a marketing research methodology that has been rarely utilised by academics: mystery-shopping. Mystery-shopping has the potential to revolutionise marketing research in the context of e-commerce.

The research tentatively asserts that prices, time elapsed since launch, parent brand association, service quality and security have an influence on consumer purchase intentions. It can be argued that unlike in traditional banking services where functional elements are more important in influencing purchase intentions, technical qualities may be more important in influencing purchase intentions in Internet banking services. This does not mean that functional qualities are not important, but Internet banking may lead to a reduction in their importance. Further, although on a theoretical level it can argued that Internet banking service products display qualities that are more akin to physical products, there is a lack of evidence to support this claim. Finally, once the novelty wears off, Internet banking services may become to be perceived as 'traditional', or more specifically the present distinction may diminish, with consumers receiving banking services through both physical and Internet channels.

References

Adrianssens, C. and Cadman, L. (1999), An Adaptation of Moderated e-mail Focus Groups to Assess the Potential for a New Online (Internet) Financial Services Offer in the UK, *Journal of Market Research Society,* 41(4), pp. 417-424

Akerlof, G. (1970), The Market for Lemons: Quality Uncertainty and the Market Mechanism, *Quarterly Journal of Economics,* 84 (August), pp. 488-500

Anderson, E. W. Fornell, C. and Lehmann, D. (1994), Customer Satisfaction, Market Share and Profitability: Findings from Sweden, *Journal of Marketing,* 58 (July), pp. 53-66

Anderson, E. W. and Fornell, C. (1994), A Customer Satisfaction Research Prospectus in Theory and Practice, in Rust, R. T. and Oliver, R. L., (Eds.), *Service Quality: New Directions,* pp. 241-268, Sage, London.

Angur, M.G., Nataraajan, R. and Jahera Jr., J. S. (1999), Service Quality in the Banking Industry: An Assessment in a Developing Economy, *International Journal of Bank Marketing,* 17(3), pp. 116-123

Armstrong, J. and Arvantis, C. (1996), Mystery Shopping: Evolution of Large Scale Programmes with Financial Institutions, *ESOMAR/EFMA Conference on Banking and Insurance Optimising Challenge,* Barcelona.

Arora, R., Cavusgil, T. S. and Nevin, J. R. (1985), Evaluation of Financial Institutions by Bank versus Savings and Loan Customers: An Analysis of Factor Congruency, *The International Journal of Bank Marketing,* 3(3), pp. 47-55

Ashforth, B. and Humphrey, R. (1993), Emotional Labor in Service Roles: The Influence on Identity, in Blake E., (Ed.), *Academy of Management Review,* 18(1), pp. 88-116

Athanassopoulos, A. D. (1997), Another Look Into the Agenda of Customer Satisfaction: Focusing on Service Providers' Own and Perceived Viewpoints, *International Journal of Bank Marketing,* 15(7), pp. 264-278

Avkiran, N. K. (1994), Developing an Instrument to Measure Customer Service Quality in Branch Banking, *International Journal of Bank Marketing,* 12(6), pp.10-18

Avkiran, N. K. (1999), Quality Customer Service Demands Human Contact, *International Journal of Bank Marketing,* 17(2), pp.61-71

Babakus, E. and Boller, G.W. (1992), An Empirical Assessment of the SERVQUAL Scale, *Journal of Business Research,* 24, pp. 253-268

Bachelor, L. (2000), Can You bank on the Future?, *Moneywise,* (October), pp. 43-48 Baker, M.J. (1983), *Market Development,* Penguin, Harmondsworth.

Banking.Com (2000), *http://www.iii.co.uk/onlinebanking/?type=discussion,* [Accessed between 01/02/2000 - 01/04/2000]

Barwise, P. and Robertson, T. (1992), Brand Portfolios, *Europe Management Journal,* 10(3), pp. 7-27

Bateson, J. (1977), Do We Need Service Marketing?, *Marketing Consumer Services: New Insights,* Report 77-115, Marketing Science Institute, Boston.

Bateson, J. (1985), Perceived Control and the Service Encounter, in Czepiel, J., Solomon, M. and Surprenant, C. (Eds.), *The Service Encounter,* Lexington Books, Lexington, Mass.

Bateson, J. (1995), *Managing Services Marketing,* 3rd Edition, Dryden, Fort Worh Bauer, C. and Scharl, A. (2000), Quantitative Evaluation of Web Site Content and Structure, *Internet Research: Electronic Networking Applications and Policy,* 10(1), pp. 31-43

Bearden, W. 0. and Teel, J.E. (1983), Selected Determinants of Consumer Satisfaction and Complaint Reports, *Journal of Marketing Research,* 20(February), pp. 21- 28.

Bearden, W. 0., Lichtenstein, D.R. and Teel, J.E. (1984), Comparison Price, Coupon, and Brand Effects on Consumer Reactions to Retail Newspaper Advertisements, *Journal of Retailing,* 60 (Summer), pp. 11-36.

Bednar, D.A, Reeves, C. A. and Lawrence, R. C. (1995), The Role of Technology in Banking: Listen to the Customer, *Journal of Retail Banking,* 17(3), pp. 35-41

Bell, H. and Tang, N.K.H (1998), The Effectiveness of Commercial Internet Web Sites: A User Perspective, *Internet Research: Electronic Networking Applications and Policy,* 8(3), pp. 219-228

Bell, W. (1981), Tactical Service Marketing and the Process of Remixing, pp. 163-167, in Donnelley, J. and George, W. (Eds.), *Marketing of Services,* American Marketing Association, AMA, Chicago.

Bemers-Lee, T., Cailliau, R., Luotonen, A., Nielsen, H. F. and Secret, A. (1994), The World Wide Web, *Communications of the ACM,* 37(8), pp. 76-82

Berry, L. (1980), Services Marketing is Different, *Business,* 30(May-June), pp. 24-29 Berry, L.L. (1983), Relationship Marketing, in Berry, L.L., Shostack, G. L. and Upah, G. D. (Eds.), *Emerging Perspectives in Services Marketing,* AMA, Chicago.

Berry, L.L., Parasuraman, A. and Zeithaml, V. (1988), The Service Quality Puzzle, *Business Horizons,* (Sept/Oct), pp. 35-43

Berry, L.L., Zeithaml, V. and Parasuraman, A. (1985), Quality Counts in Services Too, *Business Horizons,* (May/June), pp. 44-52

Berst, J. (1998), Seven Deadly Web Site Sins (And Why You Must Avoid Them at All Costs), *http://www.zdnet.com/anchordesk/story/story _l 716.html,* [Accessed 06/02/1998]

Bettman, J. (1975), Issues in Designing Consumer Information Environments, *Journal of Consumer Research,* 2(December), pp. 169-177

Bettman, J. (1979), *An Information Processing Theory of Consumer Choice,* Addison Wesley, Mass.

Bharadwaj, S.G., Varadarajan, P.R. and Fahy, J. (1993), Sustainable Competitive Advantage in Service Industries: A Conceptual Model and Research Propositions, *Journal of Marketing,* 57(0ctober), pp. 83-99.

Bidault, F. and Nihtila, J. (1999), Sensitivities of Shared Product Development, *Mastering Information Management, Financial Times,* (22nd February), pp. 11- 12

Biswas, A. and Blair, E.A. (1991), Contextual Effects of Reference Prices in Retail Advertisements, *Journal of Marketing,* 55(Spring), pp. 1-12.

Biswas, A., Wilson, E.J. and Licata, J.J. (1993), Reference Pricing Studies in Marketing: A Synthesis of Research Results, *Journal of Business Research,* 27(3), pp. 239- 56.

Bitner, M. J. (1990), Evaluating Service Encounters: The Effects of Physical Surrounding on Employee Responses, *Journal of Marketing,* 54(2), pp. 69-82.

Bitner, M. J. (1992), Servicescapes: The Impact of Physical Surroundings on Customers and Employees, *Journal of Marketing,* 56, pp. 57-71

Bitner, M. J. and Hubbert, A.R. (1994), Encounter Satisfaction versus Overall Satisfaction versus Quality in Theory and Practice, in Rust, R. T. and Oliver, R. L., (Eds.), *Senlice Quality: New Directions,* pp. 72-94, Sage, London.

Bitner, M., Booms, B. and Tetrault, M. (1990), The Service Encounter: Diagnosing Favourable and Unfavourable Incidents, *Journal of Marketing,* 54(January), pp. 71-84

Black, N., Lockett, A., Ennew, C. and Winklhofee, H. (1999), The Adoption of Internet Financial Services: A Qualitative Study, *Working Paper Series,* University of Nottingham Business School.

Bloemer, J. Ruyter, K. de and Peeters, P. (1998), Investigating Drivers of Bank Loyalty: The Complex Relationship Between Image, Service Quality and Satisfaction, *International Journal of Bank Marketing,* 16(7), pp.276-286

Blower, J. (1998), How to Create a Deadly Web Site, *http://www.interbiznet.com/nomad/archives/98012 6.html,* [Accessed 12/04/1998]

Blumer, H. (1953), Psychological Import of the Human Group, in Sherif, M. and Wilson, M. (Eds.), *Group Relations at the Crossroads,* pp. 185-202, Harper and Row, NY.

Bolton, RN. and Drew, J.H. (1991a), A Multistage Model of Customers' Assessments of Service Quality and Value, *Journal of Consumer Research,* 17(March), pp. 375-384

Bolton, RN. and Drew, J.H. (1991b), A Longitudinal Analysis of the Impact of Service Changes on Customer Attitudes, *Journal of Marketing,* 55(January), pp. 1-9

Boulding, W., Kalra, A., Staelin, R and Zeithaml, V.A. (1993), A Dynamic Process Model of Service Quality: From Expectations to Behavioural Intentions, *Journal of Marketing Research,* 30(February), pp. 7-27

Brealey, R and Myres, S. (1984), *Principles of Corporate Finance,* McGraw-Hill, Tokyo.

Bremser, W. (1997), Making Frames Fit, *Internet World,* September, pp. 90-92 Brown, C., Sopp, L. and Gould, S. (1997), Consumers' Association Goes Undercover, *Consumer Policy Review,* 7(1), pp. 2-6

Brown, S. and Swartz, T. (1989), A Gap Analysis of Professional Service Quality, *Journal of Marketing,* 53(April), pp. 92-98.

Brown, T.,J., Churchill, Jr., G. A. and Peter, J. P. (1993), Research Note: Improving The Measurement of Service Quality, *Journal of Retailing,* 69(1), pp. 127-139

Brucks, M. and Schurr, P. (1990), The Effects of Bargainable Attributes and Attribute Range Knowledge on Consumer Choice Process, *Journal of Consumer Research,* 16(March), pp. 409-419

Bucklin, L. (1996), Testing Propensities to Shop, *Journal of Marketing,* 30(January), pp. 22-27

Burnside, A. (1994), In Store Spies Snuff Out Poor Service, *Marketing,* 28(April), pp. 32-33

Busch, D. (1997), Avoid the Five Cardinal Graphical Sins, *Internet World,* (June), pp. 98-99

Carman, J. and Uhl, K. (1973), *Marketing: Principles and Methods,* Irwin, Homewood, Ill.

Carman, J.M. (1990), Consumer Perceptions of Service Quality: An Assessment of the SERVQUAL Dimensions, *Journal of Retailing,* 66(1), pp. 33-55

Chaffey, D., Mayer, R, Johnston, K and Ellis-Chadwick, F. (2000), *Internet Marketing,* Prentice hall, Harlow.

Churchill, G.A. Jr. and Surprenant, C. (1982), An Investigation into the Determinants of Customer Satisfaction, *Journal of Marketing Research,* 19(November), pp. 491- 504

COI (Central Office of Information) (1996), *Vehicle Inspectorate, Traffic Safety and a Cleaner Environment: A Report on the Vehicle Inspectorate's Effectiveness as an Enforcement Agency,* 1995/96

Cowell, D. (1984), *The Marketing of Services,* Heinemann, London.

Coyne, K. P. and Dye, R. (1998), The Competitive Dynamics of Network Based Business, *Harvard Business Review,* 76(1), pp. 99-109.

Cramp, B. (1994), Witnesses for the Defense, *Marketing,* 18(August), p. 18 Cronbach, L.J, and Meehl, P. (1955), Construct Validity in Psychological Tests, *Psychological Bulletin,* 52, pp. 281-302

Cronbach, L.J, Rajaratnam, N. and Glesser, G. C. (1963), A Theory of Generalizability: A Liberation of Reliability Theory, *British Journal of Statistical Psychology,* 16, pp. 137-163

Cronin Jr., J.J. and Taylor, S.A. (1992), Measuring Service Quality - A Re-examination and Extension, *Journal of Marketing,* 56(July), pp. 55-68

Cronin Jr., J.J. and Taylor, S.A. (1994), SERVPERF Versus SERVQUAL: Reconciling Performance Based and Perceptions Minus Expectations Measurement of Service Quality, *Journal of Marketing,* 58(January), pp. 125-131

Crosby, P. B. (1979), *Quality is Free: The Art of Making Quality Certain,* New American Library, NY.

Curwin, J. and Slalater, R. (1996), *Quantitative Methods for Business Decisions,* 4th Edition, Thompson Business Press, London.

Czepiel, J., Solomon, M. and Surprenant, C. (1985), *The Service Encounter,* Lexington Books, Lexington, Mass.

Daniel, E. and Storey, C. (1997), On line Banking: Strategic and Management Challenges, *Long Range Planning,* 30(6), pp. 890-898

Dannesnberg, M. and Kellner, D. (1998), The Bank of Tomorrow with Today's Technology, *International Journal of Bank Marketing,* 16(2), pp. 90-97

Darby, M. R. and Kami, E. (1973), Free Competition and the Optimal Amount of Fraud, *Journal of Law and Economics,* 16(April), pp. 67-86

Datamonitor (1999 various), *http://www.datamonitor.com/,* London.

Davis, F. D. (1989), Perceived Usefulness, Perceived Ease of Use, and User Acceptance of Information Technology, *MIS Quarterly,* 13(3), pp. 319-340

Dawson, J. and Hilliar, J. (1995), Competitor Mystery Shopping: Methodological Considerations and Implications for the MRS Code of Conduct, *Journal of the Market Research Society,* (October), pp. 417-427

Day, A (1997), A Model for Monitoring Web Site Effectiveness, *Internet Research: Electronic Networking Applications and Policy,* 7(2), pp. 109-115

Day, RS. (1984), Toward a Process Model of Consumer Satisfaction, in *Conceptualization and Measurement of Consumer satisfaction and Dissatisfaction,* Hunt, H. (Ed.), Marketing Science Institute, Cambridge, MA.

Della, B., Monroe, A. K.B. and McGinnis, J.M. (1981), Consumer Perceptions of Comparative Price Advertisements, *Journal of Marketing Research,* 18 (November), pp. 416-427.

Dem, D. P. (1994), Myth or Menace? A History of Business on the Net, *Internet World,* (Jul/Aug), pp. 80-91

Devlin, J.F. and Ennew, C.T. (1997), Understanding Competitive Advantage: The Case of Financial Services, *International Journal of Bank Marketing,* 15(3), pp. 73- 82.

Devlin, J.F., Ennew, C.T. and Mirza, M. (1995), Organisational Positioning in Financial Services Retailing, *Journal of Marketing Management,* 11, pp. 119-132.

Diacon, S.R. and Ennew, C. T. (1996), Ethical Issues in Insurance Marketing in the UK, *European Journal of Marketing,* 30(5), pp. 67-80

Dodds, W.B. and Monroe, K.B. (1985), The effect of brand and price information on subjective product evaluations, in Hirschman, E.C. and Holbrook, M.B. (Eds.), *Advances in Consumer Research,* Vol. 12, Association for Consumer Research, Provo, UT, pp. 85-90.

Dodds, W.B., Monroe, K.B. and Grewal, D. (1991), Effects of Price, Brand and Store Information on Buyers' Product Evaluations, *Journal of Marketing Research,* 28(August), pp. 307-319

Dodson, J. A., Tybout, A.M. and Stemthal, B. (1978), Impact of Deals and Deal Retraction on Brand Switching, *Journal of Marketing Research,* 15 (February), pp. 72-81.

Doll, W.J. and Torkzadeh, G. (1988), The Measurement of End-User Computing Satisfaction, *MIS Quarterly,* 12, pp. 259-274

Doll, W.J. Xia, W. and Torkzadeh, G. (1994), A Confirmatory Factor Analysis of the End-User Computing Satisfaction Instrument, *MIS Quarterly,* 18, pp. 453-461

Dommeyer, C. J. and Moriarty, E. (1999), Comparing Two Forms of E-mail Survey: Embedded vs. Attached, *Journal of the Market Research Society,* 41, pp. 39-50.

Doob, A. N., J. 1 Carlsmith, J.M, Freedman, J.L., Landauer, T.K. and Soleng, T., Jr. (1969), Effect of Initial Selling Price on Subsequent Sales, *Journal of Personality and Social Psychology,* 11(4), pp. 345-50.

Dorman, K. G. (1994), Results Can Shape Your Future, *Bank Marketing,* (August), pp. 17-21

Downes, L. and Mui, C. (1998), *Killer App - Digital Strategies for Market Dominance,* Harvard Business School Press, Boston, USA.

Drake, C., Gwynne, A. and Waite, N. (1998), Barclays Life Customer Satisfaction and Loyalty Tracking Survey: A Demonstration of Customer Loyalty Research in Practise, *International Journal of Bank Marketing,* 16(7), pp. 287-292

Dwek, R. (1996), Magic of Mystery Shopping, *Marketing,* 17(0ctober), pp. 41-44

Economist (1999), Various, *The Economist,* The Economist Newspaper, London, also available at *http://www.economist.com/*

Economist (2000), Various, *The Economist,* The Economist Newspaper, London, also available at *http://www.economist.com/* egg (2000), Press Release, *http://www.egg.com,* [Accessed 21/09/2000].

Eiglier, P. *et al.* (1977), *Marketing Consumer Services: New Insights,* Marketing Sciences Institute, Cambridge, MA.

Engel, J., Kollatt, D. and Blackwell, P. (1968), *Consumer Behaviour,* 2nd Edition Dryden, New York.

Engel, J., Kollatt, D. and Blackwell, P. (1991), *Consumer Behaviour,* 6th Edition Dryden, New York.

Enis, B.M. and Roering, K.J. (1981), Services Marketing: Different Products, Similar Strategy, in Donnelley, J. and George, W. (Eds.), *Marketing of Services,* AMA, Chicago.

Ennew, C. and McKechnie, S. (1998), The Financial Services Consumer, in Gabbot, M. and Hogg, G. (Eds.), *Consumers and Services,* pp. 187-207, John Wiley & Sons, Chishester.

Erdem, T. (1998), An Empirical Analysis of Umbrella Branding, *Journal of Marketing Research,* 35(3), pp. 339-352

Esmail, A. and Everington, S. (1993), Racial Discrimination Against Doctors From Ethnic Minorities, *British Medical Journal,* 306, 13(March), pp. 691-692

Evans, P. (1999), Strategy and the New Economics of Information, *Mastering Information Management, Financial Times,* (8th February), pp. 3-4

Evered, R. and Louis, M.R. (1981), Alternative Perspectives in the Organizational Sciences: "Inquiry From the Inside" And "Inquiry From the Outside", *Academy of Management Review,* 6(3), pp. 385-395

Farquhar, P. H. (1989), Managing Brand Equity, *Marketing Research,* 1(September), pp. 24-33.

Fink, A. and Kosecoff, J. (1985), *How to Conduct Surveys - A Step By Step Guide,* Sage Publications, London.

Fletcher, K. and Hastings, W. (1983), The Relevance of the Fishbein Model to Insurance Buying, *Services Industries Journal,* 3(3), pp. 296-307

Foley, P. and Jayawardhena, C. (1999), Internet Data Sources and Surveys, *European Business Review,* 11(2), pp. 105-108

Fornell, C. (1991), A National Customer Satisfaction Barometer: The Swedish Experience, *Journal of Marketing,* 56, pp. 6-21

Forrester (1998 to 2000 various), *http://wwwforrester.com/Home/0,3257,2,FF.html,* Cambridge, Massachusetts.

Friedrichs, J. and Ludtke, H. (1975), *Participant Observation,* Lexington, MA, USA Furlong, C.B. and Ritchie, J.R.B. (1986), Consumer Concept Testing of Personal Financial Services, *The International Journal of Bank Marketing,* 4(1), pp. 3-18

Gabbott, M. (1996), Don't Just Sit There - Stand Up and Talk to Me - Understanding Interpersonal Orientation and Service Technology, Proceedings of the, *American Marketing Association Winter Educators Conference,* Hilton Head, pp. 321-326

Gabbott, M., and Hogg, G (1994), Consumer Behaviour and Services: A Review, *Journal of Marketing Management,* 10(4), pp. 311-24.

Gabbott, M., and Hogg, G (1998), *Consumers and Services,* John Wiley and Sons, Chichester.

Gandy, A. (1996), Stand and Deliver: IT's Threat, *Chartered Banker,* (December), pp. 22-23

Gandy, A. (1998a), How the Internet Will Shape Global Banking, *Chartered Banker,* (April), pp. 32-33

Gandy, A. (1998b), Smart Solutions for Banking Supremacy, *Chartered Banker,* (July), pp. 41-44

Gandy, A. (1998c), Channel Delivery Strategies and Customer Relationship Management, *Chartered Banker,* (October), pp. 22-61

Gandy, A. and Brierly, C. (1997), Special Report: The Impact of Internet Technology, *Chartered Banker,* (September), pp. 11-42

Gandy, A. and Chapman, C. (1996), *The Electronic Bank - Banking and IT in Partnership,* The Chartered Institute of Bankers, Canterbury, Kent.

Gandy, A. and Chapman, C. (1997), Special Report: Data Warehousing, *Chartered Banker,* (May), pp. 14-41

Gardyne, A (1998), Internet Success Stories, *http://www.aia.net.au/success,* [Accessed 12/12/1998]

Garvin, D. A. (1983), Quality on the Line, *Harvard Business Review,* 61(Sept-Oct), pp. 65-73

Gehrke, D. and Turban, E. (1999), Determinants of Successful Website Design: Relative Importance and Recommendations for Effectiveness, Proceedings of the *32nd Hawaii International Conference on System Sciences - 1999,* Honalulu, Hawaii.

Gentle, C. J. S. (1993), *The Financial Services Industry,* Avebury, Rants.

Giga Information group (1999 various), *http://www.gigaweb.com/,* Cambridge, Massachusetts.

Gomez.com (2000), Internet Banking Scorecard, *http://www.gomez.com/channels/index. cfm?topcat_id=J,* [Accessed between *0110212000 - 0110412000]*

Graham, G. (1996), Essential Banking for the 21st Century, *Financial Times,* (15th October), p.19

Grether, D. and Wilde, L. (1984), An Analysis of Consumer Choice: Theory and Experiments, *Journal of Consumer Research,* 10(March), pp. 373-385

Grewal, D., Krishnan, R., Baker, J. and Borin, N. (1998a), The Effect of Store Name, Brand Name and Price Discounts on Consumers' Evaluations and Purchase Intentions, *Journal of Retailing,* 74(3), pp. 331-352

Grewal, D., Monroe, K.B. and Krishnan, R. (1998b), The Effects of Price-comparison on Buyers Perceptions of Acquisition Value, Transaction Value, and Behavioural Intentions, *Journal of Marketing,* 62(2), pp. 46-59

Gronroos, C. (1978), A Service Oriented Approach to Marketing of Services, *European Journal of Marketing,* 12(8), pp. 588-601

Gronroos, C. (1981), Internal Marketing - An Integral Part of Marketing Theory, in Donnelly, J.H. and George, W.R. (Eds.), *Marketing of Services,* AMA, Chicago.

Gronroos, C. (1982), *A Service Quality Model and Its Management Implications,* Research Report, EIASM Workshop on Research into the Management of Services, London.

Gronroos, C. (1984), A Service Quality Model and its Marketing Implications, *European Journal of Marketing,* 18(4), pp. 36-44

Gronroos, C. (1988), Service Quality: The Six Criteria for Good Perceived Service Quality, *Review of Business,* 9(3), pp. 10-13

Gronroos, C. (1990), *Service Management and Marketing: Managing the Moments of Truth in Service Competition,* Lexington, MA.

Gronroos, C. (1993), Toward a Third Phase in Service Quality Research: Challenges and Future Directions, in Teresa A. Swartz, David E. Bowen, and Stephen W. Brown (Eds.), *Advances in Services Marketing and Management: Research and Practice,* Vol. 2, pp. 49-64, JAI Press, Greenwich, CT.

Grove, S.J. and Fisk, R. (1992), Observational Data Collection Methods for Services Marketing: An Overview, *Journal of the Academy of Marketing Science,* 20(3), pp. 217-24

Gummesson, E. (1987), The New Marketing: Developing Long-Term Interactive Relationships, *Long Range Planning,* 20(4), pp. 10-20

GVU (2000), GVUs User Surveys http://www.cc.gatech.edu/gvu/user_surveys/User_Survey_Home.html, [Accessed between *0110212000 - 0110412000*]

Hagel III, J., Hewlin, T. and Hutchings, T. (1997), Retail Banking: Caught in a Web, *The Mckinsey Quarterly,* 2, pp. 42-55

Haine, D. (1998), Five Most Serious Web Design Errors, *E-Business (http://hpcc920.external.hp.com/Ebusiness/webdesign. html),* [Accessed 03/05/1998]

Hair, J. F. Jr., Anderson, R. E., Tatham, R. L. and Black, W. C. (1995), *Multivariate Data Analysis,* 4th Edition, Prentice Hall, Upper Saddle River, New Jersey.

Hamilton, A. (1997), Avoid the #1 Website Sin: Slow Loading Pages, *http://www4.zdnet.com/anchordesk/story/story 1244.html,* [Accessed 22/09/1997]

Heaney, J. G. and Goldsmith, R. E. (1999), External Information Search for Banking Services, *The International Journal of Bank Marketing,* 17(7), pp. 305-324

Heath, J. (1998), Pointers on How to Create Websites That Work, *http://www.iinet.net.au/-heath,* [Accessed 12/03/1998]

Henderson, R. (1995), European Retail Banking: Innovation Strategies, *International Journal of Business Studies,* 3 (1), pp. 11-30

Herr, P. M. and Fazio, R.H. (1992), The Attitude-to-Behaviour Process: Implications for Consumer Behaviour in Andrew A. M. (Ed.), *Advertising Exposure, Memory, and Choice,* Hillsdale, NJ: Lawrence Erlbaum Associates.

Hiltz, S. and Johnson, K. (1990), User Satisfaction with Computer Mediated Communication Systems, *Management Science,* 36(6), pp. 739-764

Ho, J. (1997), Evaluating the World Wide Web: A Global Study of Commercial Sites, *Journal of Computer Mediated Communication,* 1(3), http://www. ascusc.org/jmc/vol1/issue3/ [Accessed 24/02/2000]

Hochschild, A. (1983), *The Managed Heart: Commercialisation of Human Feeling,* (University of California Press, Berkeley), referenced in Czepiel, *et al.,* (1985), op. cit.

Hoffman, D. L. and Novak, T. P. (1996a), You can't sell if you don't have a market you can count on, *Harvard Business Review,* (Nov-Dec), p. 161

Hoffman, D. L. and Novak, T. P. (1996b), Marketing in Hypermedia Computer-Mediated Environments: Conceptual Foundations, *Journal of Marketing,* 60(3), pp. 50-68

Hoffman, D. L., Novak, T.P. and Chatterjee, P. (1997), Commercial Scenarios for the Web: Opportunities and Challenges, *Journal of Computer Mediated Communication,* 1(3), http://www.ascusc.org/jmc/vol 1/issue3/ [Accessed 10/10/1999]

Hoffman, D.L. and Novak, T. P. (1996c), A New Marketing Paradigm for Electronic Commerce, Paper submitted for the Special Issue on *Electronic Commerce for the Information Society,* http://www.ecommerce.vanderbuilt.edu/novaklnew.marketing.paradigm.html [Accessed 29/09/99]

Hokansson, H. (1982), An Interaction Approach, in Hokansson, H. (Ed.), *International Marketing and Purchasing of Industrial Goods,* John Wiley and Sons, Chichester.

Howard, J.A. and Sheth, J.N. (1969), *The Theory of Buyer Behaviour,* John Wiley and Sons, New York.

Huber, C.P., Lane, K.R. and Pofcher, S (1998), Format Renewal in Banks - It's Not Easy, *The McKinsey Quarterly,* Winter, (1), pp. 148-154

Huber, J. and McCann, J. (1982), The Impact of Inferential Beliefs on Product Evaluations, *Journal of Marketing Research,* 19 (August), pp. 324-33.

IFBG Gottingen (2000), *http://www.gwdg.de/ ifbglbanking.html,* [Accessed between 01/02/2000 - 01/04/2000] iii (2000), Interactive Investor International *http://www.banking.com/,* [Accessed between 01102/2000 - 01/0412000]

Inman, J. and McAlister, L. (1993), A Retailer Promotion Policy Model Considering Promotion Signal Sensitivity, *Marketing Science,* 12(Fall), pp. 339-356

Jacoby, J. (1984), Perspectives on Information Overload, *Journal of Consumer Research,* 10(March), pp. 185-202

Jacoby, J., Chestnut, R. and Fisher, W (1978), A Behavioural Process Approach to Information Acquisition in Non-durable Purchasing, *Journal of Marketing Research,* 15, (3-August), pp. 532.544

Jain, A.K., Pinson, C. and Malhotra, N.K. (1987), Customer Loyalty as a Construct in the Marketing of Banking Services, *The International Journal of Bank Marketing,* 5(3), pp. 49-72

Janal, D. S. (1997), *Online Marketing Handbook - How to Promote, Advertise, and Sell Your Products and Services on the Internet,* John Wiley and Sons, New York.

Janal, D. S. (2000), *Dan Janal's Guide to Marketing on the Internet,* John Wiley and Sons, New York.

Jayawardhena, C. (1999a), Security and Internet Payment Systems, *European Business Review,* 11(6), pp. 357-358.

Jayawardhena, C. (1999b), Impediments to the Growth of Commercial Transactions on the Internet - Internet Payment Systems, in Jenkins, A.M. (Ed.), *Global Electronic Commerce, ISRC Working Paper 180599,* University of Baltimore, Baltimore, pp. 6-13.

Jayawardhena, C. (2000), From Measurement to Understanding Process, Proceedings of the, *e-Business Challenges - A Research Agenda,* Foundation for Management Education, Midlands Business Schools Management Research Seminars and Workshops, Aston Business School, Aston University, Birmingham, UK.

Jayawardhena, C. and Foley, P. (1998), Overcoming Constraints on Electronic Commerce - Internet Payment Systems, *Journal of General Management,* 24(2), pp. 19-35.

Jayawardhena, C. and Foley, P. (2000), Changes In Banking Sector - The Case of Internet Banking in UK, *Journal of Internet Research: Networking and Policy,* 10(1), pp. 19-30.

Jayawardhena, C.R.K. (1997), *The Role of Financial Institutions in Commercial Property Cycles with Particular Attention to the UK,* Unpublished MPhil Disseration, University of Cambridge, Cambridge.

Johnson, M. (1984b), Consumer Choice Strategies for Comparing Non-comparable Alternatives, *Journal of Consumer Research,* 11 (December), pp. 741-753

Johnson, T.(1984a), The Myth of Declining Brand Loyalty, *Journal of Advertising Research,* 24(February/March), pp. 9-17.

Johnston, A. (1999), Welcome to the Wired World, *Research,* (November), pp. 22-25 Johnston, M.W., Parasuraman, A., Futrell, C.M. and Black, W.C. (1990), A Longitudinal Assessment of the Impact of Selected Organisational Influences of Salespeople's Organisational Commitment During Early Employment, *Journal of Marketing Research,* 27(3), pp. 333-344

Johnston, R. (1988), Service Industries: Improving Competitive Performance, *Service Industries Journal,* 8(2), pp. 202-211

Johnston, R. (1997), Identifying The Critical Determinants of Service Quality in Retail Banking: Importance and Effect, *International Journal of Bank Marketing,* 15(4), pp. 111-116

Johnston, R., Silvertro, R., Fitzgerald, L. and Voss, C. (1990), Developing the Determinants of Service Qualtiy, in Langeard, E. and Eiglier, P. (Eds.), *Marketing, Operations and Human Resources Insights into Services,* 1st International Research Seminar on Services Management, IAE, Aix-en-Provence, pp. 373-400.

Johnston, W. and Bonama, T. (1981), Purchase Process for Capital Equipment and Services, *Industrial Marketing,* 4, pp. 253-264

Jorgensen, D. L. (1989), *Participant Observation: A Methodology for Human Studies,* Sage Publications, Newbury Park, CA.

Judd, T. (1964), The Case for Redefining Services, *Journal of Marketing,* 28(January), pp. 58-59

Jupitor (1999 various), *http://wwwjup.com/home jsp,* New York.

Kalyanaram, G. and Winer, R.S. (1995), Empirical Generalizations From Reference Price Research, *Marketing Science,* 14(3), pp. 161-169.

Kamen, J. and Toman, R. (1970), Psychographics of Prices, *Journal of Marketing Research,* 8(1), pp. 252-261

Keller, K. and Staelin, R. (1987), Effects of Quality and Quantity of Information on Decision Effectiveness, *Journal of Consumer Research,* 14(September), pp.200- 213

Kelly, S.W., Donnelly, J. H. and Skinner, S. J. (1990), Customer Participation in Service Production and Delivery, *Journal of Retailing,* 66(3), pp. 315-335

Kent, R. and Lee, M. (1999), Using the Internet for Market Research, *Journal of Market Research Society,* 41(4), pp. 377-385

Khazeth, K. and Decker, W. H. (1992), How Customers Choose Banks, *Journal of Retail Banking,* 14(4), pp. 41-44

Kim, P. (1990), A Perspective on Brands, *Journal of Consumer Marketing,* 7(Fall), pp. 63-67.

Kirmani, A. and Wright, P. (1989), Money Talks: Perceived Advertising Expense and Expected Product Quality, *Journal of Consumer Research,* 16(December), pp. 344-353.

Knights, D., Sturdy, A. and Morgan, D. (1994), The Consumer Rules: An Examination of Rhetoric and "Reality" of Marketing in Financial Services, *European Journal of Marketing,* 28(3), pp. 42-54

Kotler, P. (1972), *Marketing Management: Analysis, Planning, Implementation & Control,* 2nd Edition, Prentice-Hall, Englewood Cliffs, NJ. Kotler, P. (1982), *Principles of Marketing,* Prentice Hall, New Jersey Kotler, P. (1991), *A1arketing Management,* Prentice Hall, New Jersey.

Kotler, P. and Bloom, P. (1984), *Marketing Professional Services,* Prentice Hall, Englewood Cliffs, NJ.

Kotler, P. and Levy, S.J. (1969), Broadening the Concept of Marketing, *Journal of Marketing,* 33, pp.10-15

Kramer, D. (1997), In a Flash Tutorial: Adding Life to Your Web Pages with Micromedia Flash, *Internet Voyager,* 3(6), pp. 16-18

Krishnan, H. S. and Chakravarti, D. (1993), Varieties of Brand Memory Induced by Advertising: Determinants, Measures, and Relationships in David A. A. and Alexander L. B., (Eds.), *Brand Equity and Advertising,* Hillsdale, NJ: Lawrence Erlbaum Associates.

LaBarbera, P. A. and Mazursky, D. (1983), A Longitudinal Assessment of Consumer Satisfaction/Dissatisfaction: The Dynamic Aspect of the Cognitive Process, *Journal of Marketing Research,* 20(November), pp. 393-404.

Lane, C. and Bachman, R. (1996), The Social Construction of Supplier Relations in Britain and Germany, *Organizational Studies,* 17(3), pp. 365-75

Langeard, E., Bateson, J., Lovelock, C. and Eiglier, P. (1981), *Marketing of Services: New Insights from Consumers and Managers,* Marketing Sciences Institute, Cambridge, MA.

Laroche, M., and Manning, T. (1984), Consumer Brand Selection and Categorisation Processes: A Study of Bank Choice, *The International Journal of Bank Marketing,* 2(3), pp. 3-21

Laroche, M., Kim, C.and Zhou, L.. (1996)., Brand Familiarity and Confidence Determinants of Purchase Intention: An Empirical Test in a Multiple Brand Context, *Journal of Business Research,* 37, pp. 115-120.

Leeds, B. (1995), Mystery Shopping: From Novelty to Necessity, *Bank Marketing,* (June), pp. 17-23

Lehman Brothers (2000 various), *http://www.lehman.com/,* New York,

Lehtinen, U. and Lehtinen, J. R. (1991), Two Approaches to Service Quality Dimensions, *Service Industries Journal,* 11(3), pp. 287-303

Leiner, B. M., Cerf, V. G., Clark, D.D., Kahn, R. E., Kleinrock, L., Lynch, D.C., Postel, J., Roberts, L.G and Wolff, S.S. (1997), The Past and Future History of the Internet, *Communications of the ACM,* 40(2), pp. 102-108

Leonard, M. and Spencer, A. (1991), The Importance of Image as a Competitive Strategy: An Exploratory Study in Commercial Banks, *The International Journal of Bank Marketing,* 9(4), pp. 25-29

Levesque, T. and McDougall, G.H.G. (1996), Determinants of Customer Satisfaction in Retail Banking, *International Journal of Bank Marketing,* 14(7), pp. 12-20

Levitt, J. (1972), Product Line Approach to Service, *Harvard Business Review,* (Sept-Oct), p. 43

Levitt, J. (1976), The Industrialisation of Services, *Harvard Business Review,* (Sept-Oct), pp. 63-74

Levitt, T. (1981a), Marketing Intangible Products and Product Intangibles, *Harvard Business Review,* 59(May-June), pp. 94-102

Levitt, T. (1981b), Product-line Approach to Service, *Harvard Business Review,* (Sept-Oct), p. 43

Lewis, B.R. (1991), Service Quality: An International Comparison of Bank Customers' Expectations and Perceptions, *Journal of Marketing Management,* 7(1), pp. 47- 62

Lewis, B.R. and Mitchell, V.M. (1990), Defining and Measuring the Quality of Customer Service, *Marketing Intelligence & Planning,* 8(6), pp. 11-17

Lewis, B.R., Orledge, J. and Mitchell, V.W. (1994), Service Quality: Student's Assessment of Banks and Building Societies, *International Journal of Bank Marketing,* 12(4), pp. 3-12

Light, L. (1990), , *How Advertising and Promotion Help to Build Brand Assets,* speech before the Second Annual Advertising Research Foundation Advertising and Promotion Workshop, (February), pp. 22-23.

Llewellyn, D. (1996), Banking in the 21st Century: The Transformation of an Industry, *Loughborough University Banking Centre - Research Paper,* Loughborough University, Loughborough.

Locander, W. and Hermann, P. (1979), The Effect of Self-confidence and Anxiety of Information Seeking in Consumer Risk Reduction, *Journal of Marketing Research,* 19, pp. 268-274

Lockett, A. and Littler, D. (1997), The Adoption of Direct Banking Services, *Sloan Management Review,* 13, pp. 791-811

Lovelock, C. H. (1988), *Managing Services,* Prentice-Hall, Englewood Cliffs. Lowe, A. and Kuusisto, J. (1999), The Institutional Stature of the Retail Bank: The Neglected Asset, *International Journal of Bank Marketing,* 17(4), pp. 171-181

Lumby, S. (1994), *Investment Appraisal & Financing Decisions,* VNR, London Mandeville, T. (1998), An Information Economics Perspective on Innovation, *International Journal of Social Economics,* 25(2/3/4), pp. 357-364

Marshall, J. (1998), Electronic Banking: Top Trio Emerges, *USBanker,* 180(2), p. 18 Martenson, R. (1985), Consumer Choice Selection in Retail Bank Selection, *The International Journal of Bank Marketing,* 3(2), pp. 64-74

Martin, P. (1998a), The Net's Loss, *Financial Times,* (20th October), p. 20 McKechnie, S. (1992), Consumer Buying Behaviour in Financial Services: An Overview, *International Journal of Bank Marketing,* 10(5), pp.5-12

McKenzie, G. and Khalidi, M. (1996), Globalization of Banking and Financial Markets: The Challenge for European Regulators, *Journal of European Public Policy,* 3(4), pp. 629-46

McLuhan, M. (1964), *Understanding Media,* McGraw Hill, New York.

McNeil, R. (1994), The Mystery Squad's Tougher Challenge, *Research Plus,* (April), p. 13

Mehta, R. and Sivada, E. (1995), Comparing Response Rates and Response Content in Mail Versus Electronic Mail Surveys, *Journal of Market Research Society,* 37(4), pp. 429-439

Miles, L. (1993), Rise of the Mystery Shopper, *Marketing,* July, (29), pp.19-20 Mols, N. P. (1998a), The Internet and the Banks' Strategic Distribution Channel Decisions, *Internet Research: Electronic Networking Applications and Policy,* 8(4), pp. 331-337

Mols, N.P (1998b), The Behavioural Consequences of PC Banking, *International Journal of Bank Marketing,* 16(5), pp. 195-201

Monroe, K. (1979), , *Pricing: Making Profitable Decisions, McGraw-Hill Book Company, New York, NY.*

Monroe, K. B., Grewal, D. and Compeau, L.D.. (1991), The Concept of Reference Prices: Theoretical Justifications and Research Issues, *Working paper, University of Illinois, Champaign-Urbana, IL.,* University of Illinois, Champaign-Urbana, IL.

Monroe, K.B. and Chapman, J.D. (1987), Framing Effects on Buyers' Subjective Product Evaluations, in Wallendorf, M. and Anderson, P. (Eds.), *Advances in Consumer Research,* Vol. 14, Association for Consumer Research, Provo, UT, pp. 193-197.

Morrison, L.J., Colman, A. M. and Preston, C. C. (1997), Mystery Customer Research: Cognitive Processes Affecting Accuracy, *Journal of the Market Research Society,* 39(2), pp. 349-361

Moschandreas, M. (1994), Business Economics, *Routledge, London.,*

Murray, K. (1991), A Test of Services Marketing Theory: Consumer Information Acquisition Activities, *Journal of Marketing,* 55(January), pp.10-25

Muylle, S., Moenaert, R. and Despontin, M. (1999), Introducing Website User Satisfaction: An Integration of a Qualitative Pilot Study with Related MIS Research, *Working Paper,* Owen Graduate School of Management, Vanderbuilt University, Nashville, TN.

Nachmias, C.F. and Nachmias (1993), *Research Methods in the Social Sciences,* 4th Edition, Edward Arnold, London.

Nagle, T. (1982), Economic Foundations for Pricing, *Pricing Strategy,* University of Rochester (September 24-25), USA.

Nellis, J. (1998), Strategies For Staying Ahead, *Chartered Banker,* (June), pp.28-31 Nelson, P. (1970), Information and Consumer Behaviour, *Journal of Political Economy,* 78 (2), 311-29.

Nelson, P. (1974), Advertising as Information, *Journal of Political Economy,* 82 (July/August), pp. 729-754.

Newman, K. and Cowling, A. (1996), Service Quality in Retail Banking: The Experience of Two British Clearing Banks, *International Journal of Bank Marketing,* 14(6), pp. 3-11

Newman, K., Cowling, A. and Leigh, S. (1998), Case Study: Service Quality, Business Process Re-engineering and Human Resources: A Case in Point?, *International Journal of Bank Marketing,* 16(6), pp. 225-243

Newton, E. (1994), The Service Murders by Miss Terri Shopper, *Research Plus*, (September), p. 4

Nicosia, F. (1966), *Consumer Decision Process: Marketing and Advertising Implications*, Prentice Hall, New Jersey.

Nunley, K. (1998), Promoting Your Biz Online: Get Noticed, *Wealth Building*, (March), p. 47

Oliver, R. L. (1980), A Cognitive Model of the Antecedents and Consequences of Satisfaction Decisions, *Journal of Marketing Research*, 17(November), pp. 460- 469.

Oliver, R. L. and Swan, J.E. (1989), Consumer Perceptions of Interpersonal Equity and Satisfaction in Transactions: A Field Survey Approach, *Journal of Marketing*, 53(April), pp. 21-35.

Oliver, R.L. (1980), A Cognitive Model of the Antecedents and Consequents of Satisfaction Decisions, *Journal of Marketing Research*, l 7(November), pp.460- 469

Olson, J. C. and Jacoby, J.. (1977)., Consumer Response to Price: An Attitudinal, Information Processing Perspective in Wind, Y. and G. Grenberg, G. M, (Eds.), *Moving Ahead With Attitude Research*, pp. 73-86, American Marketing Association, Chicago.

Ostrow, J. (1998), MSFDC Puts Financial Institutions at the Heart of Internet Bill Presentment and Payment, *Journal of Retail Banking*, 20(3), pp. 1-7

Palmer, A. (1998), *Principles of Services Marketing*, McGraw-Hill, Berkshire. Parasuraman, A., Zeithaml, V.A. and Berry, L. (1985), A Conceptual Model of Service Quality and its Implications for Future Research, *Journal of Marketing*, 49(Fall), pp. 41-50

Parasuraman, A., Zeithaml, V.A. and Berry, L. (1988), SERVQUAL: A Multiple-Item Scale for Measuring Consumer Perceptions of Service Quality, *Journal of Retailing*, 64(1), pp. 12-40

Parasuraman, A., Zeithaml, V.A. and Berry, L. (1991), Refinement and Reassessment of the SERVQUAL scale, *Journal of Retailing*, 67(Winter), pp. 420-450

Parasuraman, A., Zeithaml, V.A. and Berry, L. (1994a), Reassessment of Expectations as a Comparison Standard in Measuring Service Quality: Implications for Further Research, *Journal of Marketing*, 58(January), pp. 114-124

Parasuraman, A., Zeithaml, V.A. and Berry, L. (1994b), Alternative Scales for Measuring Service Quality: A Comparative Assessment Based on Psychometric and Diagnostic Criteria, *Journal of Retailing*, 70(3), pp. 201-230

Parkin, M. and King, D. (1995), *Economics,* Addison-Wesley, Workingham, England

Peterson, M. and Rajan, R. (1995), The Benefits of Lending Relationships, *Quarterly Journal of Economics,* 110, pp. 407-443

Pincott, G. and Branthwaite, A. (2000), Nothing New Under the Sun?, *International Journal of Market Research,* 42(2), pp. 137-155

Pine II, B.J. (1993), *Mass Customization: The New Frontier in Business Competition,* Harvard Business School Press, Boston, MA.

Pine II, J. H. and Gilmore, J. H. (1998), Welcome to the Experience Economy, *Harvard Business Review,* (July-August), pp. 97-107

Prescott, M.B. and Van Slyke, C. (1997), Understanding the Internet as an Innovation, *Industrial Management and Data Systems,* 97/3, pp. 119-124

Price., L. L., Amould, E. J. and Tierney, P. (1995), Going to Extremes: Managing Service Encounters and Assessing Provider Performance, *Journal of Marketing,* 49(Fall), pp. 25-45

Quelch, J. A. and Klein, L. R. (1996), The Internet and International Marketing, *Sloan Management Review,* 37(3), pp. 60-75

Quelch, J. and Ash, S. (1991), Consumer Satisfaction with Professional Services, in Donnelly, J. and George, W. (Eds.), *Marketing of Services,* pp. 82-85, AMA, Chicago.

Rajan, R.G. (1998), The Past and Future of Commercial Banking - Viewed through an Incomplete Contract Lens, *Journal of Money, Credit, and Banking,* 30(3), pp. 524-550

Rao, A. R. and Monroe, K.B. (1989)., The Moderating Effect of Prior Knowledge on Cue Utilization in Product Evaluations, *Journal of Consumer Research,* 15 (September), pp. 253-264.

Rao, A. R. and Sieben, W. (1992)., The Effect of Prior Knowledge on Price Acceptability and the Type of Information Examined, *Journal of Consumer Research,* 19 (September), pp. 256-270.

Rathmell, J. M. (1974), *Marketing in the Service Sector,* Winthrop Publishers Inc., Cambridge, Mass.

Read, S. (1998), Online Banking, *The Guardian,* October 31, pp. I-N

Reed, K. and Miles, L.R. (1995), How Mystery Shoppers Support 'The Brenton Way' of Creating and Defining a Service Culture, *Journal of Retail Banking Services,* Winter, 17(4), pp. 21-28

Regan, W. (1963), The Service Revolution, *Journal of Marketing,* 27(July), pp. 57-62 Richardson, P.S., Dick. A. S., and Jain, A.K (1994)., Extrinsic and

Intrinsic Cue Effects on Perceptions of Store Brand Quality, *Journal of Marketing,* 58(0ctober), pp. 28-30.

Robertson, T. S. (1970), *Innovative Behaviour and Communications,* Holt-Rheinhart, NY.

Rogerson, S, Foley, P. and Jayawardhena, C. (1999), Is Electronic Commerce a Socially Beneficial Activity? in Armitage, J. and Roberts, J. (Eds.), *Exploring Cyber Society,* Newcastle.

Rust, R. T. and Oliver, R. L. (1994), *Service Quality: New Directions in Theory and Practice,* Sage, London.

Rust, R., Zahonik, A. and Keningham, T. (1996), *Service Marketing,* Harper Collins, NY.

Rust, R.T., Zahorik, J. and Keiningham, T.L. (1995), Return on Quality (ROQ): Making Service Quality Financially Accountable, *Journal of Marketing,* 59(April), pp. 58-70

Ryan, B. (1991), *It Works! How Investment Spending in Advertising Pays Off,* American Association of Advertising Agencies, New York.

Sagawa, P. (1997), The Balkanization of the internet, *The McKinsey Quarterly,* 1, pp. 126-139

Sasser, W., Olson, R. and Wyckoff, D. (1978), *Management of Service Operations,* Allyn & Bacon, Boston, MA.

Schillewaert, N., Langerak, F. and Duhamel, T. (1998), Non-probability Sampling for WWW Surveys: A Comparison of Methods, *Journal of Market Research Society,* 40(4), pp. 307-322

Scholl, N., Mulders, S. and Drent, R. (1999), Online Qualitative Market Research: Interviewing the World at a Fingertip, Proceedings from the, *Esomar Qualitative Conference,* Athens, Greece.

Segal-Hom, S. (1988), Global Service Delivery - Managing the Critical Interdependencies, *Proceedings of the Operations Management Association Conference,* Warwick, pp. 99-128

Selltiz, C., Wrightsman, L.S. and Cook, S.W. (1976), *Research Methods in Social Relations,* 3rd Edition, NY: Holt, Rinehart and Winston.

Sharpe, S. A. (1990), Asymmetric Information, Bank Lending, and Implicit Contracts: A Sylised Model of Customer Relationships, *The Journal of Finance,* 45(4), pp. 1069-1087

Shostack, G. L (1982), How to Design a Service, *European Journal of Marketing,* 16(1), pp. 49-63

Shostack, L (1977), Breaking Free From Product Marketing, *Journal of Marketing,* 41(April), pp. 73-80

Smith, V. (1997), What Every Good Webmaster Knows, *Home Office Computing,* (October), pp. 56-58

Spence, M. (1974), *Market Signalling,* Cambridge, MA: Harvard University Press.

Statt, D. (1997), *Understanding the Consumer,* MacMillan Press, Basingstoke.

Stem, B., Gould, S. and Tewari, S. (1993), Sex-typed Service Images: An Empirical Investigation of Self-Service Variables, *The Service Industries Journal,* 13(3), pp. 74-96

Sterne, J. (1996), *Customer Service on the Internet,* John Wiley and Sons, New York.

Sterne, J. (1999), *World Wide Web Marketing - Integrating the Web into Your Marketing Strategy,* 2°d Edition, John Wiley and Sons, New York.

Sterne, J. and Priore, A. (2000), *Email Marketing - Using Email to Reach Your Target Audience and Build Customer Relationships,* John Wiley and Sons, New York.

Stigler, G. (1961), The Economics of Information, *Journal of Political Economy,* 69(June), pp. 213-25.

Strauss, J. and Frost, R. (1999), *Marketing on the Internet,* Prentice hall, Upper Saddle River, NJ.

Strong, E. K. (1925), *The Psychology of Selling,* McGraw-Hill, NY.

Surprenant, C. and Solomon, M. (1987), Predictability and Personalisation in the Service Encounter, *Journal of Marketing,* 51(March), pp. 86-96

Swan, J.E. and Combs, L. J. (1976), Product Performance and Consumer Satisfaction: A New Concept, *Journal of Marketing,* (April), pp. 45-62

Tadjer, R. (1998), How to Avoid the 10 Fatal Website Mistakes, *http://www.techweb.com/netbiz/archives/fatalmistakesstop1-2.html.* [Accessed 22/03/1998]

Taylor, P. (1998), Internet Set for 5% of Sales, *Financial Times,* (10th November), p.7 Teas, R.K. (1993), Expectations, Performance Evaluation, and Consumer Perceptions of Quality, *Journal of Marketing,* 57(0ctober), pp. 18-34

Teas, R.K. (1994), Expectations as a Comparison Standard in Measuring Service Quality: An Assessment of a Reassessment, *Journal of Marketing,* 58(January), pp. 132-139

Thaler, R. (1985), Mental Accounting and Consumer Choice, *Marketing Science,* 4(Summer), pp. 199-214

Thomas, D. (1978), Strategy is Different in Service Business, *Harvard Business Review,* (July), pp. 158-165

UK Internet Banking.com (2000), *http://www.ukinternetbanking.com/,* (Accessed between 01102/2000 - 01/04/2000]

Urbany, J. And Weilbaker, D. (1987), A Critical Examination of Nelson's Theory of Information and Consumer Behaviour, in Douglas, S. *et al.,* (Eds.), *AMA Educators Conference Proceedings,* AMA, Chicago.

Warburg Dillion Read (1999 various), *http://www.ubs.com/,* Zurich, Switzerland Weber, R.P. (1985), *Basic Content Analysis,* Sage Publications, California, USA.

Wilson, A and Gutman, J. (1998), Public Transport: The Role of Mystery Shopping in Investment Decisions, *Journal of the Market Research Society,* Oct, 40(4), pp.285-95

Wilson, A. (1972), *The Marketing of Professional Services,* McGraw Hill, Maindenhead Winer, R. S. (1986), A Reference Price Model of Brand Choice for Frequently Purchased Products, *Journal of Consumer Research,* 13(September), pp. 250-256.

Wison, A. (1998), The Use of Mystery Shopping in the Measurement of Service Delivery, *Service Industries Journal,* 18(3), pp. 148-163

Woodruff, R.B., Cadotte, E.R. and Jenkins, R.L. (1983), Modeling Consumer Satisfaction Process Using Experience Based Norms, *Journal of Marketing Research,* 20(August), pp. 296-304

Woodside, A. G., Frey, L. L. and Robert, T. D. (1989), Linking Service Quality, Customer Satisfaction, and Behavioural Intentions, *Journal of Health Care Marketing,* 9, pp. 5-17.

Wright, L.T. and Jayawardhena, C. (2001), Netting the Consumer: The E-Direct Marketing Imperative, Proceedings of the Tenth Biennial World Marketing Congress, *Global Marketing Issues at the Turn of the Millennium,* Cardiff, Wales, UK, (forthcoming)

Wylie, I. (1999), Online, But on the Case?, *Guardian,* (23rd January), p.23 Yin, R.K. (1993), *Applications of Case Study Research,* Sage, London.

Yin, R.K. (1994), *Case Study Research - Design and Methods,* Sage, London.

Zakon, R.H. (1998), Hobbes' Internet Timeline v3.3, *http://info.isoc.orgl-zakon/internet/History/HIT html,* [Accessed 17/11/98]

ZD Net UK (2000), *http://www.zdnet.co.uk/pcmag/labs/1998/12/online-banking/*, [Accessed between 01/02/2000 - 01/04/2000]

Zeithaml, V. A. (1988), Consumer Perceptions of Price, Quality and Value: A Means-End Model and Synthesis of Evidence, *Journal of Marketing,* 52 (July), pp. 2- 22.

Zeithaml, V. A. and Graham, K. L. (1983), The Accuracy of Reported Reference Prices for Professional Services in Bagozzi, R. P. and Tybouts, A. M., (Eds.), *Advances in Consumer Research,* Vol. 10, pp. 607-611, Association for Consumer Research.

Zeithaml, V. and Bitner, M. (1996), *Services Marketing,* McGraw Hill, New York, NY Zeithaml, V. Gilly, M. (1987), Characteristics Affecting the Acceptance of Retailing Technologies: A Comparison of Elderly and Non Elderly Consumers, *Journal of Retailing,* 63 (Spring), pp. 49-68

Zeithaml, V.A. (1981), How Consumer Evaluation Process Differ Between Goods and Services in Donnolly, J. and George, W., (Eds.), *The Marketing of Services,,* pp. 186-190, American Marketing Association, Chicago. 4anything (2000), 4anything - Survey on Internet Banking", *http://4net-banks.4anything.com/4/0,1001,5126,00.html,* [Accessed between 01/02/2000 - 01/04/2000]

Appendix A

The following is a brief account along with a Web screen snapshot of the different Internet banking services in the research.

1. BankNet Electronic Banking Service

URL — http://www.mkn.co.uk/bank

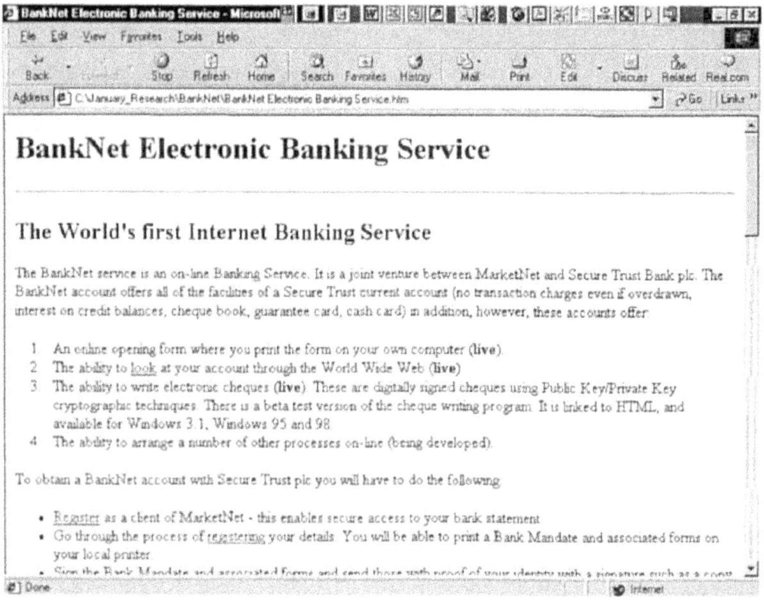

BankNet is a stand-alone Internet banking service. It MarketNet and Secure Trust Bank plc. Secure Trust plc, formerly called the People's Bank, has been in existence for more than 90 years. It is now a part of the MarketNet, and the Internet banking service run by them is called BankNet. Secure Trust Bank subscribes to the Code of Banking Practice and is an authorised institution under the Banking Act, 1987.

The Web site is the smallest (in terms of the number of pages, the amount of information, etc.) and the simplest of all bank Web sites in the analysis. All the Web pages are written in simple black text on white background. There is virtually no graphics on the site, save for a banner at the bottom of each page. The Web site cannot be synchronized, but if the user visits the Web site previously, it does work offline using pages from the temporary folder, provider the user has enabled the feature. Some of the links do not work indicating that there are Web authoring deficiencies and lack of regular updates.

To obtain a BankNet account with Secure Trust plc a consumer will have to carry out the following:-

a) Register as a client of MarketNet - this enables secure access to the bank statement.

b) Go through the process of registering the details. It will enable the consumer to print a Bank Mandate and associated forms on a local printer.

c) Sign the Bank Mandate and associated forms and send those with proof of identity with a signature such as a copy of the Passport or Drivers' Licence (this is a requirement of the money laundering regulations).

d) Download the Internet WorkHorse program. This will enable the consumer to write electronic cheques. It is only available for Windows 3.1. The current release does write electronic cheques.

It is difficult to see why the developers of this Internet banking service appear to develop applications that do not work on the most popular and current operating systems such as Windows 95 or Windows 98. Windows 3.1 is an operating system that has been gradually been phased out from the beginning of 1995. Only a very small number of consumers would still be using Windows 3.1, and they too would only be using it if their computers are fairly dated machines, unable to handle more recent operating systems. Therefore it appears that the promoters of this service are not in tune with the present day developments and cater to a very small portion of prospective customers. Further BankNet is not usable by all users running their computers on other operating platforms, such as Unix, Linux, Apple, etc.

Any version of Netscape can be used to look up accounts. Elsewhere on the Web site, it states that the WorkHorse - only runs under Windows and Windows95. This appears to be a clear contradiction of what is stated on the home page.

When sending confidential information it is conducted over a secure server.

The last update on the site appears to be in April 1999, which in Internet terms is a very long time.

1.1 How an Electronic Cheque Works

The BankNet EPI or ECheque allows account holders to instruct Secure Trust Bank to pay funds out of their account to another individual within the network. The ECheque is only one of a number of signed instructions that can be used for commercial purposes. Current instructions include:

- Electronic Payment Instruction - ECheque
- Investment Instruction - buy or sell shares
- Certificate Request - for confirmation of identity
- Electronic Instruction Authority - to set up an electronic instruction
- Air Flight booking - to request an air flight booking.

Digitally Signed Instructions use public key cryptography to generate a digital signature. This signature essentially proves that a particular person in possession of a particular private key created the original instruction. The key is generated by the individual who wishes to use it and there is no need for that key to ever go out of the hands of its owner. The Bank, Stockbroker or Travel Agent dealing with requests on the basis of the digital instruction never need to see the private key. Nor does anyone else.

In order to carry out an electronic payment, a consumer has to follow the following steps.

1. Get a program capable of writing digitally signed instructions. 'WorkHorse' is one such program.
2. Use that program to generate a private key. BankNet recommend generating a key of 512 bits, but up to 1024 can be used.
3. The decision to what to use the key for is left to the consumer. The consumer needs to set limits as to what can be instructed with the key. For instance
    - The key will have a limited life (perhaps 6 months, perhaps shorter).
    - The key will have a limit on weekly turnover (perhaps a maximum of £100 pounds a week)
    - The key will have a limit on transaction size (perhaps £25) The Bank of England have requested that we limit transactions initially to £50.
    - The key will have a total turnover limit (perhaps £1000).
4. Register the key. This is done for Electronic Cheques by using hyperlinks found at the Web site.

5. Finally, the key can be used to instruct people to do things. It is important to remember that using the key is the same as signing a paper document or using ATMs. The consumer is committing himself or herself to a contract up to the limit of the key. If the contract is to pay for something then the consumer is agreeing to do so. The key may be limited, but the consumer is committed when they sign that key up to the amount of the instruction or the limit whichever is the lesser. It is also important to note that the consumer cannot use their key for more than the amount they have limited it to.

1.2 Demo and the Operation of the Account

There is no demonstration program or application. Operation of the account appears to be on a very basic HTML environment, with black text on white background. It offers limited functionality.

1.3 Query/Email

No email number given, a generic Web inquiry form is available on the Web site. The connection to this Web inquiry form is via an unsecured connection, and therefore is an unsuitable method of sending confidential communication.

2. First-E

URL - http://www.first-e.com/uk/index.php3

First-e is the only bank to operate throughout the European Union - UK, France, Germany and Spain. Under the European Union rules it is possible to for an organisation with a banking license from one first-e the Internet bank is not currently a member of the UK Banking Ombudsman Scheme or a signatory to the British Bankers Association (BBA), Banking Code of Practice. At present, the legislative and constitutional framework establishing these schemes do not allow first-e the Internet bank to obtain membership to them. Clearly, the 1987 UK Banking Act was framed long before the concept of Internet banking was envisaged as a means of offering consumers the best available service. In a recently published report into banking in the UK commissioned by the Government and produced by Don Cruickshank (now chairman of the London Stock Exchange), it was recommended that internet banks such as first-e, should be admitted as members of the new Financial Services Ombudsman Scheme on a voluntary basis.

It is possible to synchronize the Web site and work offline. Requires additional software for optimum performance of the site, such as ActiveX controls and plug-ins. If the browser is set to medium or higher security levels, is not safe run ActiveX controls found on this Web site. It requires the user's computer to accept cookies. In order to observe animation on the site, the computer must have Macromedia flash 4.0 or later installed. Further, the Web site uses Java extensively. There is no statement of the kind of software and hardware needed for prospective customers, and it is not explicit on as to whether the system works on different platforms.

The Web site appears to be continually updated. There appears to be a concerted effort to provide as many financials service and products in order to increase stickiness and increase cross selling. The site index and the directory structure are very effectively displayed, thereby letting the user can understand instantly where they are within the Web site. The customer comments section is biased, listing a selection of customers with extremely favourable comments. Only the customer log hyperlink opens a new window. As previously alluded to, the site design is such that the need to open new Windows is almost eliminated.

2.1 Demo and the Operation of the Account

The demo runs on a Java script.

When a customer logs-in for the first time, a large number of Java applets is downloaded and these are retained in the computer. Subsequently a customer must download an applet in order to login to their accounts. Hence it does not enable offline operation even up to the account login process. If a customer does commence the login process without having established the connection to the Internet, they will be taken to the customer login page, but they will not be able to establish the connection to the servers at first-e. The reason for this is that the system requires a customer to download the applet on each login. This is because it is likely that the applet changes weekly to reflect new facilities that are being launched by the bank. Downloading the applet upon login ensures that the customer have access to these new features. It is the intention of first-e that once all additional features have been launched, the rate at which the applets change will slow down and a new system will be brought into operation. The proposed system will check the applet during the system check and only prompt a download if there is a newer version available.

If a customer wishes to uninstall these downloaded components, it is not possible to download them using the Windows uninstall facility, rather they have to delete the files on an individual basis from the respective directories. This process is potentially hazardous, since some of these files reside in the registry directory of the operating system. If incorrect files are deleted from the registry directory it could lead to a total collapse of the operating system, making the computer unworkable until such time the operating system has been re-installed.

There is online help within the account. The account interface is similar to Windows interface and is user friendly.

In addition to the username and the passwords, a customer also needs to assign a 'TAN' for each transaction. A TAN list is a list of code numbers sent to every customer and issued in sequence or at random, in combination with their password or personal identification number. This gives each transaction a unique code. If a customer enters an incorrect password and username combination or TAN three times, that account will be soft locked for security reasons. A soft locked account can only be unlocked by telephoning the call centre.

2.2 Query/Email

Query is not communicated over a secure server. first-e attempts to answer email queries within 48 hours. Claim to achieve a target rate of 98% within 24 hours.

3. Smile

URL - http://www.smile.co.uk/

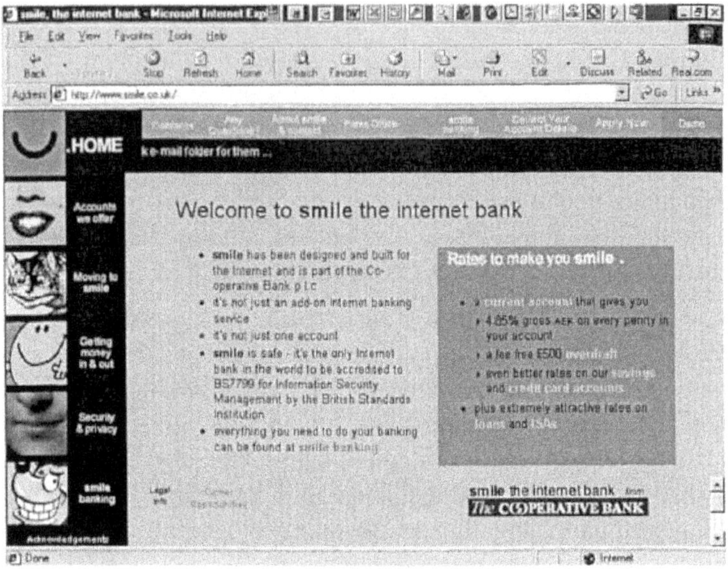

Smile is a subsidiary of the Cooperative bank, and commenced operations in October 1999.

The Web site relies on extensive use of Java. Provides wide-ranging information and is presented in very easy to understand language. Requires a PC running either a minimum IE4 or Netscape 4.06 Java-enabled browser for operation. IE5 users need to download Virtual Microsoft Machine option to be able to use smile banking and other Java programs that are necessary. smile banking does not support Macintosh users. This is because the Java coding which the system uses to ensure the security and robustness of smile banking. Unfortunately, the writers of Java have not yet found a way to make Java behave in the same way on Macs as it does on PCs. All bank charges are very clearly stated. Navigation is clear and logical, with new browser Windows opening at appropriate times.

The Web site provides very clear statements on security with firm assurances. For instance,

> *"We will repay you any money that is taken from your account due to: any error by our staff or our systems a computer crime which is not identified and stopped by our security system".*

3.1 Demo and the Operation of the Account

Demo uses flash 4. In spite of having the screen resolution set to 600x800, the recommended resolution, one cannot get the entire demo screen into one screen shot.

The account incorporates a help function. The account also has very useful features such as 'ready to use' letters. These can be used to initiate a move of an account to smile from another banks, as redirecting one's salary, get a list of existing standing orders and direct debits etc.

3.2 Query/Email

Two Internet based ways are available for contacting the bank, either through email or a Web based form.

4. Egg

URL - http://www.egg.com/

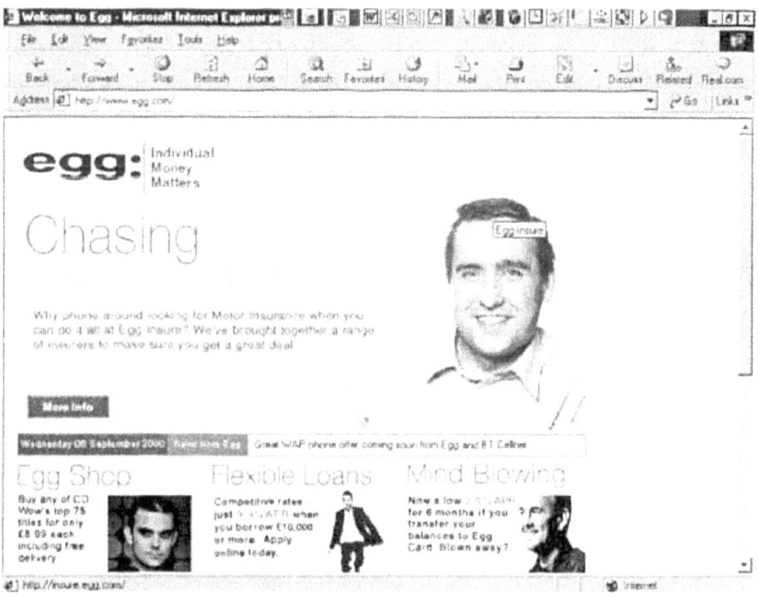

Egg was launched by the Prudential, the UK's largest insurer, in October 1998. It has had a remarkable growth since its launch. In 1999 egg took 22% of all new deposits in the British banking system (Economist, 2000). On the other hand, egg may be the only company in the world to tum to the Internet to slow down growth. It started as a bank with the telephone and the Internet as delivery channels, offering above market interest rates on savings deposits. Within six months it had reached its target of 500,000 depositors (Economist, 2000), so it stopped accepting new deposits by telephone and turned itself into dedicated Internet bank with telephone support, thereby succeeding in stemming the flow of new customers.

The Web site is configured around frames at the top and at the bottom of the screen. Cookies are in operation. The Web site is very explicit on security terminology and the systems in place. A large proportion of the site content is in 'https' (those with secure connections) areas. Timeouts are in operation over the https areas. During the browsing process, new Windows do open, at appropriate occasions. However, when a user does go into a new window zone, such as the chat zone, there is no home page icon to get back to the original pages. It is possible to encounter intermittent downloading problems, and these instances have been highlighted both in the popular press and in their chat rooms. In most occasions this is down to the inadequacy of the infrastructure in place to handle the extremely large number of users.

The site has undergone considerable changes in the recent past. It resembles an Internet portal, a discussion forum, an online shopping mall, and a site for selling a host of financial products. As with the increasing number of Internet banking services offering an ISP services, egg offers free Internet access as well. It is attempting to differentiate itself from other portals by hosting a 'funds supermarket', where a consumer may open an ISA, transfer existing PEPs or buy unit trusts and (OEICs). These financial vehicles are not egg products, rather egg is acting as a broker, selling the products at a discount on either the initial fee or from the yearly management charge or both. The investment section of the portal contains a wide range of tools to purchase, sell, monitor of ones investments. Users can for instance create virtual financial portfolios and track their investments over time.

The open discussion forum is for all Internet public to discuss, recommend, and be critical about any financial services products offered by any financial services provider, more specifically those offered by egg. Although the primary emphasis in the forum is on financial services products, it also extends to discussions on e-commerce, shopping, etc. This appears to be an attempt to build a community, and increase the frequency of user visits to the site. However participation in the discussions in the forum is rather patchy. For instance there have been gaps of almost two weeks of non activity (of contributors) from 09/08/99 (http://www.eggfreezone.com/chat.htm)

4.1 Demo and the Operation of the Account

There is no demonstration application on the Web site. Since the account displays basic HTML script connected over a secure connection it will run on most of the platforms.

The emphasis is on building a perception on the part of the customer that they are being treated as an individual. If the customer decides to telephone egg to carry out a transaction or any other service request, which could have been undertaken on the Internet, they will be charged a fee of £2. Both deposits and withdrawals can only be made by electronic transfer between an egg account and the nominated account at another UK bank or building society. Statements are issued annually on the anniversary of the original account opening. In return for these conditions customers will earn a premium interest of 0.25% over and above those customers who are able to carry out transactions over phone, deposit and withdraw money using the postal services and bank branches.

4.2 Query/Email

A Web based query is available for contacting the bank, and the connection is made over a secure server.

5. First Direct

URL - http://www.firstdirect.com/home/fs _home_ fl.html

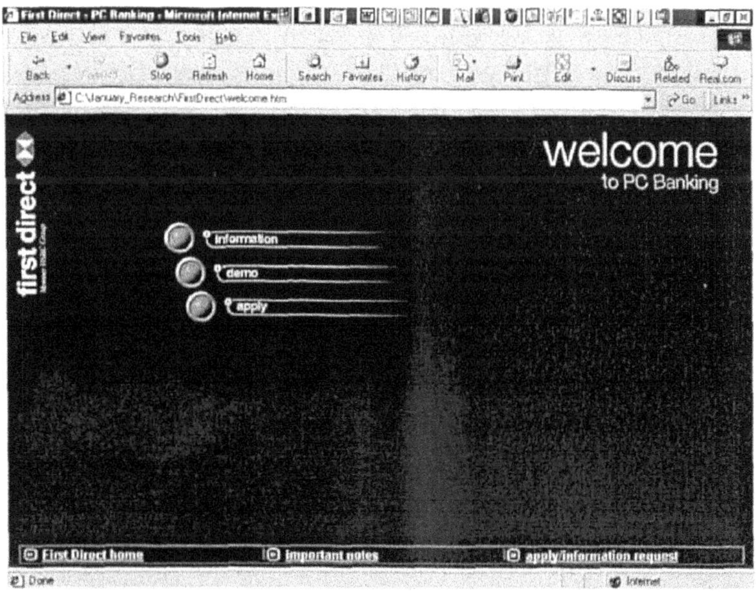

Established in 1989 by the Midland Bank, (now part of the HSBC Group), first direct pioneered telephone banking in the UK. The drive behind its inception was to provide customers with greater convenience. The

Internet banking service has been fully operational since November 1999.

The Web site of First direct is another Web site that attempts to mimic a portal with subject headings such as mail, an ISP, computers and mobile phone offers, weather, etc. However the portal services are not as extensive as the one finds in, for example with Nationwide. Search engines are made available in association with yahoo, yellow pages. The BT phone book have also been incorporated on the site which is a quite a useful feature. There is a news service with the facility for receiving regular news items from the bank.

Java is used extensively in authoring the site. The Web site relies on frames. It is not possible to synchronize and work offline. Requires additional software for optimum performance of the site, such as ActiveX controls and plug-ins, disadvantages of this is discussed above. The Web site needs cookies to operate. Additionally they need to make sure that JavaScript is enabled and that user has the latest plug-in for Macromedia Flash (Version 4.0).

Only a very little amount of information is found on a single Web page, making it necessary to go through several Web pages to find information that one needs. This process can be tedious and put off the user. The total black background with while letters is an acquired taste, but it does stand out from all the other Web sites on this design aspect. A section called the directory is used as an up market transaction site specialising in niche products like expensive chocolates, etc. A deliberate distinction is made between Internet banking site section and the product section, with each sections opening in different browser window. It appears that there are inadequate linkages between different products and how these can be managed online. The home icon may not always take the consumer to the home page of the bank.

5.1 Demo and the Operation of the Account

The demo is simple, although the site has undergone a revamp the demo has not been altered. The demo incorporates an online help feature.

It is highly reviewed by a large number of reviewers. Commercial surveys have shown that it is highly rated in terms of customer satisfaction. Customers on discussion forums have commended that transfers are usually completed in 2 instead of 5 days taken by most of the other banks.

5.2 Query/Email

Query is web based, and is not made over a secure connection. Attempts to answer emails in 7-10 working days.

6. Bank of Scotland (BOS)

URL - http://www.bankofscotland.co.uk/

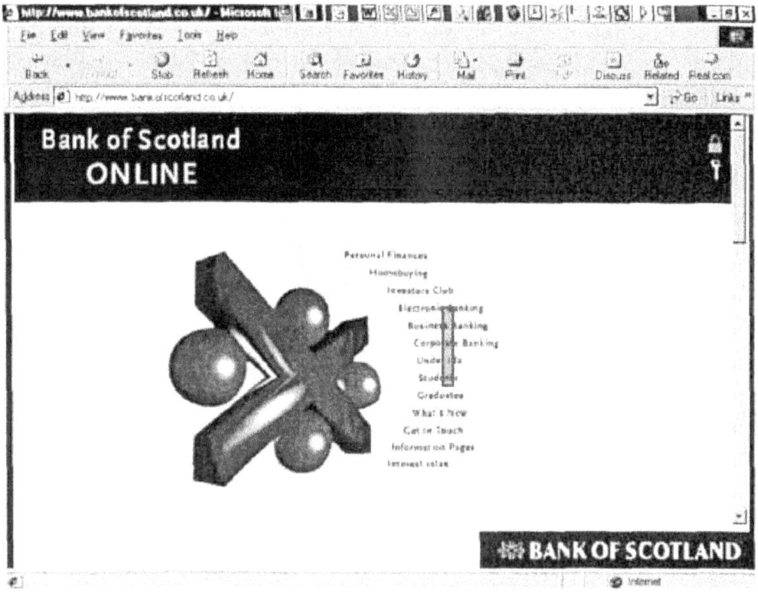

The Bank of Scotland operates a network of 325 branches, concentrated mainly in Scotland, and offers a wide range of services to business and retail customers. BOS claims that the Home and Office Banking Services (HOBS) was the first PC banking service in the UK. This PC banking service was converted into an Internet banking service in November 1999.

The central Web site incorporates an Internet portal and free Internet access in addition to the banking sections. It is not possible to synchronize the Web site and work offline. The site requires additional software such as ActiveX controls and plug-ins for optimum performance. In such a situation it may be prudent to offer a link with an option of bypassing such controls and offering the user more choice in browsing the site. The Web site needs cookies to operate. During the navigation process, it will only open a new browser window, if one leaves the entire subject area - i.e. go to the recruitment section, for example. It also treats the Internet banking section as an entirely new subject area and opens in a new window, but on the other hand all other banking product services operate on the same window. In general, Web site navigation can be difficult, there is little indication to the user informing where they are within the site. BOS does not comment as to whether HOBS runs on different computer platforms.

No pages were observed to contain a secure connection, leading to a poor overall security perception. Additionally the site lacks explicit statements about the level of security measures in place within the general areas. To read about the security features one has to download the demo. Once the consumers get to that position the explanation is good.

HOBS is only operational between 6am and lam on weekdays, until midnight at the weekend, and is therefore not a '24x7' operation as one expects an Internet banking service to be.

A consumer can only apply for Personal HOBS if they are already a Bank of Scotland customer. If they meet this requirement, all they have to do is complete the application form and send it by post.

The overall perception is that there are very few features to attract new customers.

6.1 Demo and the Operation of the Account

Demo must be downloaded as a separate file for viewing. It requires either Microsoft Internet Explorer version 4 and higher or Netscape Navigator 4.5 and higher for operation. Further it uses the Macromedia Shockwave plug in. It is advised that the demo takes an estimated download time of 3 minutes 30 seconds for 28.8k modem and about 2 minutes for 56k modems. However it takes an average download time of 12 minutes using a 56k modem.

The demo gives the user the option of running on either a self-run or user controlled mode. When the demo is operational, a large proportion of screen space from the top and bottom are taken by two frames consisting of banners, making the area available for the demo rather limited. Hence the user has to scroll down within a page to observe screen, which appears to be a rather cumbersome exercise.

A useful feature of the account interface highlighted in the demo is the messaging service. These messages refer to impending transactions on the account - so that a customer knows exactly what is going to take place in the coming week. All accounts held by a customer with the bank (that are part of the Internet banking service) can be seen at a glance. Transactions can be sorted in either ascending or descending order. A transfer (within BOS accounts or a domestic bank account) can be made to take place either immediately or up to 30 days into the future. The account interface also enables the user to order Traveler's cheques. An online help is also incorporated into the demo.

6.2 Query/Email

No email number is available, however a generic Web inquiry form is available on the Web site. The connection to this Web inquiry form is via an unsecured connection, and therefor is an unsuitable method of sending confidential information. Further there does not seem to be adequate warnings given on the form to warn users against sending confidential information through this Web form.

7. Barclays

URL - http://www.ibank.barclays.co.uk/

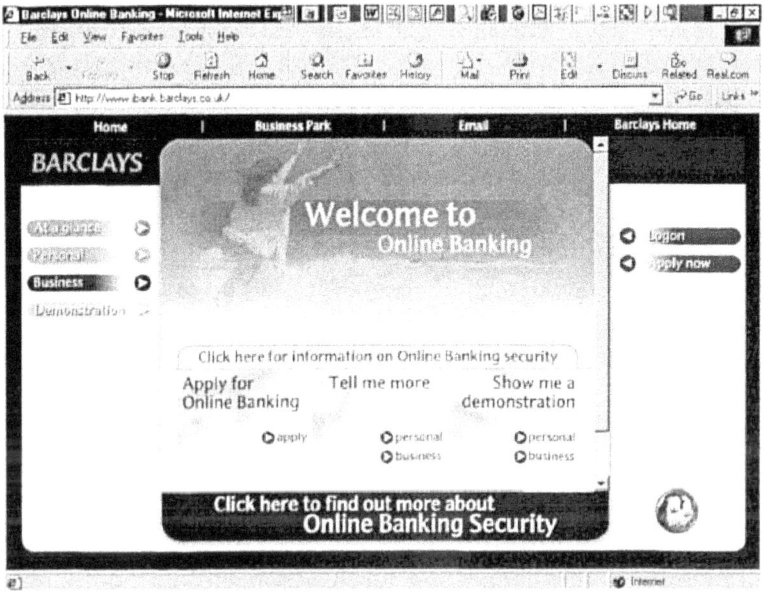

Barclays PLC is one of the largest financial services institutions in the UK and provides global coverage. The name 'Barclays' is associated with two with Freeserve, the biggest ISP in the UK. Barclays was one of the first banks to offer a free ISP service to customers. Since then many of its competitors have since followed suit. The Barclays ISP shares most content areas with Freeserve home page portal, consisting of news, chat, communities, weather, shopping, financial services, etc. thereby producing an a very prominent additional gateway to the banking site. Barclays provide some of the content, this concentrates primarily on financial services. Freeserve provides some additional content. Providing an ISP and portal content for banking customers, and non-customers, is a very prudent method of collecting additional revenue through e-commerce and advertising, and by extension gaining additional Internet exposure.

In general is can be said that this is a very well executed site. During the navigation process many new 'paths' or 'links' that open in new browser windows thereby making sure that the consumer does not get 'lost' within the site. On the other hand there is also a danger that too many browser windows open and thereby confusing the consumer. The site lacks a limiting mechanism controlling the number of new browser window openings per browsing session.

The Web site uses frames quite extensively. Graphics are fairly attractive neither too bright nor too dark. Java is also used extensively. Cookies are used when a consumer browses the Web site. Some of the icons can appear to be small depending on the screen resolution, and may present problems to the user. The text size is appropriate, considering the area dedicated for text. Most of the text is in 'bold' letters, which is helpful in making them clearly visible. It is possible to synchronize and view a large proportion of the Web site in offline mode.

The Web site lacks a search engine. The 'home page' icon can be confusing. The 'home' icon directing towards the ISP home page, and 'Barclays home' icon leading on to the banking site home page. The site download is fairly slow. The 'barclays.co.uk' home download is problematic, on several occasions browser failed to load on the page. No browser suitability function is available but the minimum requirements for using the Internet banking service are given.

Sections of the Web site alternates between secure and non-secure pages with due warning to the user. For instance the online application is submitted over a secure server. There is no separate section on security measures in place and this is surprising in light of the level of attention afforded by other Internet banking service Web sites to this topic.

7.1 Demo and the Operation of the Account

Demo opens in a new window, and consists of seven separate screens. It does not run automatically and is user controlled. Uses an identical page configuration and layout as the other areas of the Web site, thereby keeping a coherent feel about the Web site. The demo is not a separate application, and hence does not require any additional downloads.

7.2 Query/Email

There is no email facility, nor is there provision of a Web based form to communicate with the bank.

8. Citibank

URL - http://www.citibank.com/uk/intbank/index.htm

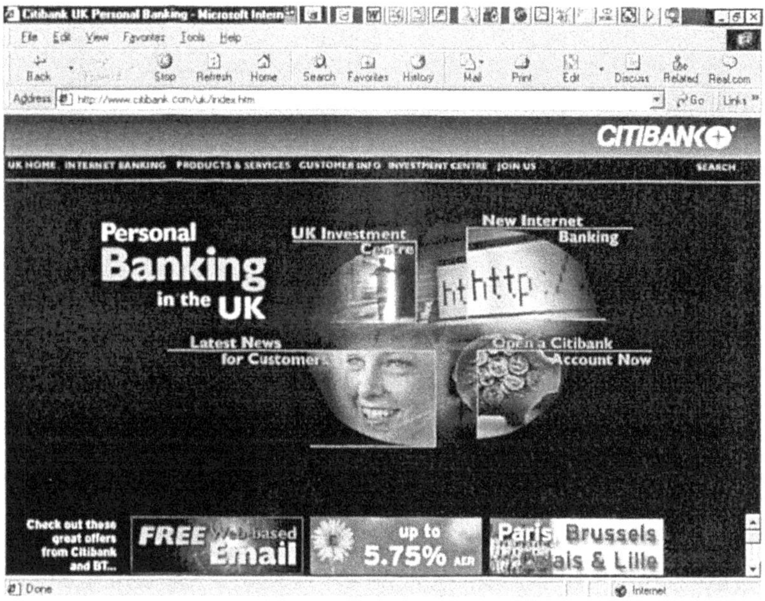

Citibank has been running its Direct Access banking service m the U.S. since 1985 via PC and now operates an Internet banking service in the UK. Citibank operates four branches in the UK, all within London but offers free withdrawals from all UK ATM machines.

Similar to the Barclays bank, Citibank offers a free ISP option in association with British Telecom (BT). The Web site contains economic and social information, an extensive section on the EMU and implications for customers. The Web site can be synchronized and made to work on offline mode. During the navigation process, new browser windows do not open. The only occasion when a new window opens is when a user fills out an online application form, in which case the connection is also made through a secure server. The Web site uses Java for messages, on a rolling basis. They also list the best screen resolution size for optimum display.

The explanation security measures employed is comprehensive and so is the FAQ section. The Web site and the accounts are best operated on Netscape Navigator/Communicator or Microsoft Internet Explorer browser 4.0 or above. The Internet banking service also supports users of AOL service.

Current Accounts may be opened in the either of three currencies, Pounds Sterling, US Dollar or Euro. This appears to be a big incentive to frequent travelers and customers who may get payments in currencies other than Pounds Sterling. Further the ability to use ATMs round the world and also to the facility to telephone free of charge from a large number of countries around the world are also positive attributes of this Internet banking service.

Prevailing Interest rates are not given on the Web site. There is provision to recommend a friend, and in the process secure a financial rewarded.

8.1 Demo and the Operation of the Account

The demo can be self-run or user controlled. The demo interface is fairly simple and it opens in a new browser window. The screen 'shots' are pasted on a Web page, with captions explaining what each of the sections are supposed to mean. It is a fairly simple in appearance, but nevertheless effective. The demo gives step-by-step instructions for downloading account information into the customer's financial management software.

Up to 99-bill payment names can be held with the account. Unlike other banks, to open an account one needs a deposit of £2000 and an annual income of at least £20,000. Personal Citibank Direct Access Identification Number (DAIN). DAIN must be entered every time an account holder login to Citibank Direct Access. An automatic time-out will occur if there is no activity for 4 minutes.

8.2 Query/Email

No email facility. It is impossible to communicate with the bank using the Internet.

9. Co-Operative Bank

URL - http://www.co-operativebank.co.uk/internet_banking.html

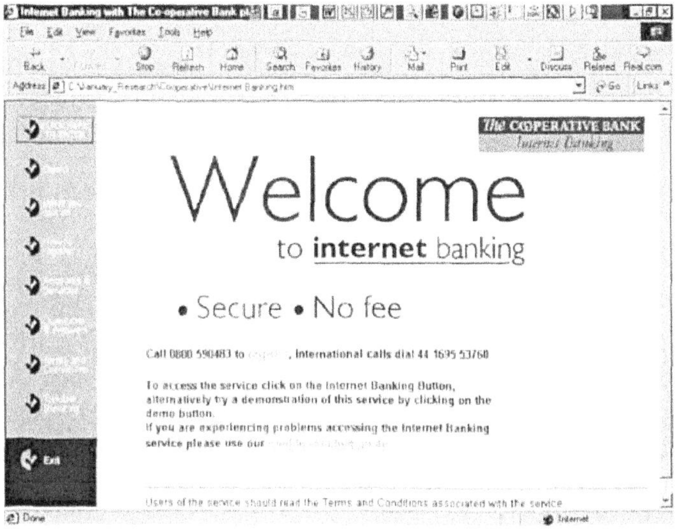

The Cooperative bank's Web site uses Java extensively in its operation. Therefore, in theory, it can run in a wide range of devices, and is not restricted to PCs, but any device supporting elements of Java, i.e. WebTV, Digital TV, personal digital assistants, WAP phones, etc. Since there are no defined standards, and often only small number of customer demand for many of these devices, the system has not been developed specifically for these operating environments. However, the Cooperative bank claim that it is possible to run their service on any Java based services running in several non-PC devices in its present form. In addition to running on IBM compatible systems, it supports some of the Macintosh systems. Earlier (pre 1998) Macintosh systems do run the service, but will not run on later Mac systems with the G3 specification or MacOS 8.

The navigation process is at times confusing. Some links result in a new browser window. At the same time, the original browser window may also open a new page with new content. However the user will be unaware that the original page has opened a new page unless they click on an icon to go back to the original page. This navigation process delivers a very confusing and cumbersome browsing experience.

Apart from the FAQ section there is a technical troubleshooting guide, both these sections are fairly comprehensive.

The Co-operative Bank is also claiming to be the UK's first interactive text service The Sky Intertext is an interactive text service that is available nationwide, and all Sky viewers can make use of this service. The only requirement of this service is to have a Teletext TV and a touch-tone telephone. With the TV remote control and the telephone a customer can control their accounts. At present the following facilities are available on this integrated service.

a) An up-to-date statement of the account on the TV screen.

b) Check balances at any time.

c) Can observe which payments and credits have been processed.

9.1 Demo and the Operation of the Account

Demo needs to be downloaded as a separate file. The mouse movements activate the downloading process. It is a Java program. The demo contains an online help feature that is very useful to users, but does not work offline. Unlike most of the demos, which merely provide screen shots of a fictitious account, this demo works just like as if a customer is operating their account. Timeout errors can take place in loading of the demo. Once this happens it is difficult to get the demo working again.

A help feature is incorporated in the account interface. Access to accounts is controlled by the following security measures,

(1) a 4-digit security number,

(2) a Java 1.1 browser (for example Netscape Navigator 4.04 or Microsoft Internet Explorer 4),

(3) 5 pieces of secure personal information.

The first time the customer use Internet Banking they download a number of Java applets into the computer cache.

The account service is not available between Monday to Saturday 05.00am to 05.30am Sunday 00.00am to 06.00am, and hence is not a '24x7' hour service.

9.2 Query/Email

An email link is available on the site, but the connection is not made over a secure server.

10. Halifax Online

URL - http://www.halifax.co.uk/

Halifax, formerly a building society, claims to be the country's number-one mortgage lender supporting more mortgage and savings accounts customers than any other organisation.

It is possible to synchronize this Web site and work offline. Requires additional software for optimum performance of the site, such as ActiveX controls and plug-ins. For optimal viewing of this site you will need Macromedia Flash version 3 or above. It requires the user's computer to accept cookies. For users wishing to avoid these attributes, or for those with 'old' browsers, there is a text only version of the Web site, which is a very useful feature. The Web site directory structure is user friendly, and attempts to indicate to the user where they are browsing.

Certain sections of the Web site are connected under a secure server connection. However, even non-confidential information flows are also connected via this 'https' server, for example even the online calculator is also on a https server. This is not only unnecessary but also adds to time lags in browsing, since there is more encrypting and decrypting involved. All this takes up more processing capacity compared to a non-secure server connection. The Web site information is not explicit on whether the system supports different computer platforms and operating systems.

10.1 Demo and the Operation of the Account

The demo takes the form of screen 'shots'. Further detailed explanatory notes are presented to further explain the workings of the account. However, because of this process the demo appears to be very long and tends to contain an excessive amount of details.

There is online help within the account. Transactions can be sorted by payment type or alphabetically. Further, they can also be sorted by date in ascending or descending order. Bill payments/transfers that have been deleted can be reinstated.

10.2 Query/Email

Email link is made over a secure server. The Web based form requires a considerable amount of information to send a simple email query. Attempts to answer emails in 48 hours.

11. Lloyds TSB On-Line

URL - http://195.92.18.3/retail/

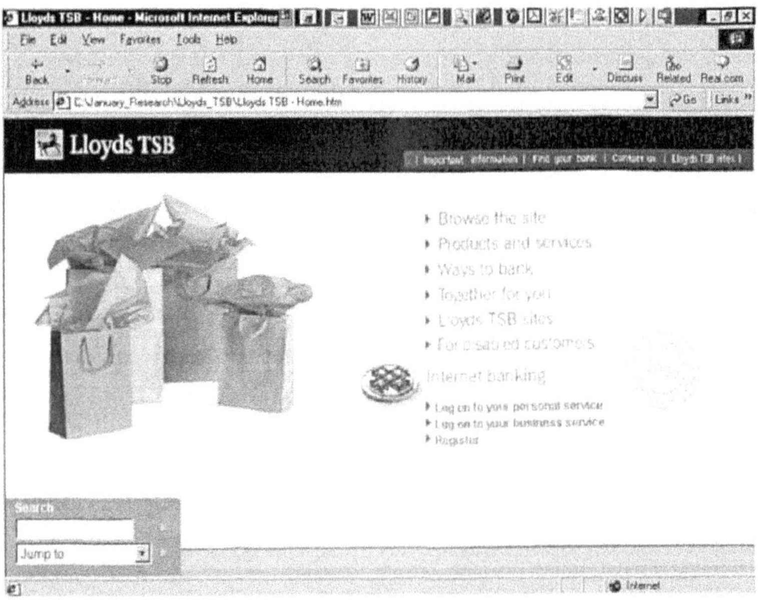

Lloyds TSB, formed by the merger of Lloyds and TSB, is one of the UK's largest banks. Integration of the two banks is far from complete but products opened with each institution before the merger can be viewed online. The site also supports products provided by Scottish Widow and loans direct, both members of the Lloyds TSB Group.

During the browsing process, the Web site requires the user's computer to accept cookies. The Web site also rely JavaScript and Java applets, but these aren't essential for the operation. A search engine is available on site. Further there is a drop down menu to take the consumer to the desired place on the Web site. There are a number of operational faults in the Web site, the most significant of all is the site map which does not appear to function. Some sections of the site are dated, as much as by a year.

New browser windows do open during the navigation process. However, when one clicks on a hyperlink in the new window (if one manually creates one), the original window changes screen without informing the user. This process along with other factors, such as changing formats for one section of the Web site to another, contribute to a rather confusing browsing experience.

The Web site is intended to act as a brochure. It is very low in interactivity and contains very few techniques for attracting non Lloyds TSB bank customers. A number of Web pages, such as those that send secure information, are connected over a secure connection.

As stated earlier, Lloyds and TSB were separate banks until their merger in the recent past. Hence there are differences among the products, two range of products are still branded separately, i.e. the Lloyds range and the TSB range. The reason for this distinction is because the two computer systems are run separately. Until total integration is, there will still be a few differences between some of the products and services, depending on whether the branch where the customer hold their account was previously called Lloyds Bank or TSB.

11.1 Demo and the Operation of the Account

Each demonstration screen 'shot' is accompanied by a small window explaining what it means, this is a quite a useful feature.

A customer must have a computer running Netscape Navigator 4 or Internet Explorer 4 or later versions to be able to use the account system. The system works both on IBM compatibles and Mac platforms. Before customers log in they have to go through a screen Netscape Certificate warning, Internet Explorer 4.5 Macintosh Edition Root Certificate Warning. This is to disable the automatic save password browser option. This appears to be a shortcoming of the service, since all the other Internet banking service providers have managed to overcome this problem.

Accounts can only be operated between 4am and midnight every day of the week, hence it is not a '24x7' Internet banking services. This is a significant disadvantage since one of the primary advantages of an Internet banking account is the ability to carry out transactions without any time restrictions whatsoever.

11.2 Query/Email

A Web based form is available for communication, but the connection is not made over a secure server.

12. Nationwide Online Banking

URL - http://www.nationwide.co.uk/

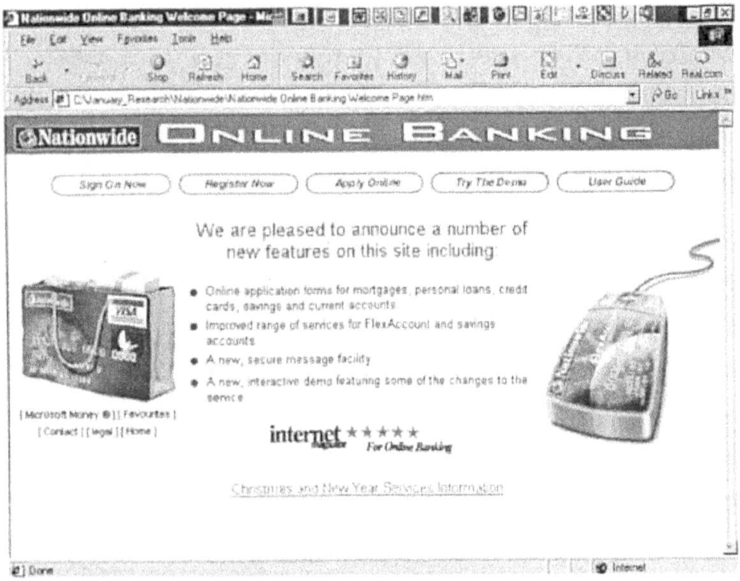

Nationwide claims to be the fourth largest mortgage lender, ninth largest retail bank and the largest building society in the UK.

It was one of the first adopters of the Internet and PC banking launching its Web site and PC Home Banking in 1995 and its Internet banking service in 1997.

Unlike a traditional building society Web site, this building society Web site is a portal with a strong emphasis on cross selling of financial products among others services and products. Other areas of the portal include, an online shopping area; a section on 'what's on', a list of entertainment scene in consumers area; a section on 'living'; links to favourite lifestyle activities of consumers such as do it yourself; ramblers; where-to-eat, etc.; other recreational facilities including 'fun & games'; and a travel section with advice and links to popular Web sites on this theme. To provide this wide range of services Nationwide have collaborated with a range of partners each with a specialisation of their own. Additionally, Nationwide provides a free ISP service.

During the browsing process, the Web site requires the user's computer to accept cookies. Frames of considerable width are found at the header and footer of each and every page. Therefore, the remaining space for other content is rather limited. A new window opens when a connection is made over a secure connection. A considerable amount of attention is paid to security, with a very detailed explanation of the security measures employed.

12.1 Demo and the Operation of the Account

Demo works on secure connection section of the Web site. Emphasis is on security with the opening screen dedicated to security. Demo is quite comprehensive. Security is assured by using three items of data.

12.2 Query/Email

Both an email link and a Web-based form can be used for communication, but the connection is not made over a secure connection.

13. NatWest Bank

URL - http://www.natwest.co.uk/frontpage/dhtml/index.htm

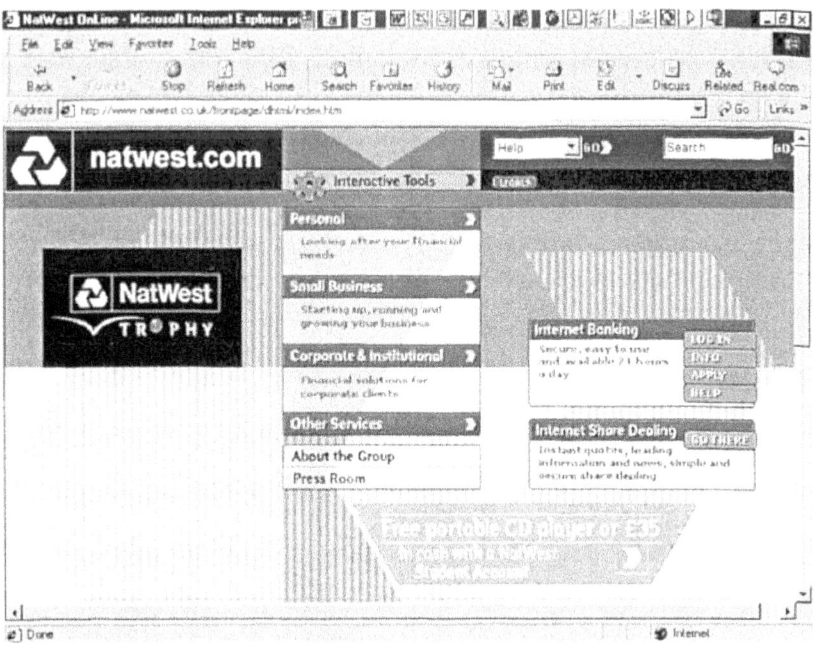

NatWest is one of the 'big four' banks in the UK. In mid 2000 it was acquired by the RBS.

It is possible to synchronize and work offline with this Web site. Requires additional software for optimum performance of the site, such as ActiveX controls and plug-ins. The site relies on extensive use of Java for animation. There is also an option for viewing the Web site without the animation features. There is a lack of consistency within the different section of the Web site, the background and the formatting theme change from one section to another. The site incorporates a search engine, however the ability of this search engine is rather limited.

There is some concern over the secure connection. On certain sections of the Web site NatWest maintain that the connection is made through a secure server but such connections do not seem to have taken place. For example, it is stated that a secure connection is made for online application, however one cannot see the secure connection being established in the form of an icon that appears in popular browsers (in IE a padlock icon is displayed on the taskbar, or in the address bar the URL begins with 'https'), nor does it prompt and inform the user that they are being taken to a location over a secure connection.

13.1 Demo and the Operation of the Account

The demo is quite effective with 10 screens. A useful feature on the account interface is a timer informing the customer they have been using the service for a certain length of time. Both IBM compatibles and Apple Macintosh platforms are supported by the system. However, elsewhere on the Web site they state that

> *"Apple Mac browser technology is currently incompatible with the leading security standards we have in place for NatWest Online Banking. However, we are currently addressing this issue and plan to offer the service to Apple Mac customers in the near future "*

13.2 Query/Email

An email link is available for communication.

14 NPBS (NetMaster)

URL - http://www.npbs.co.uk/

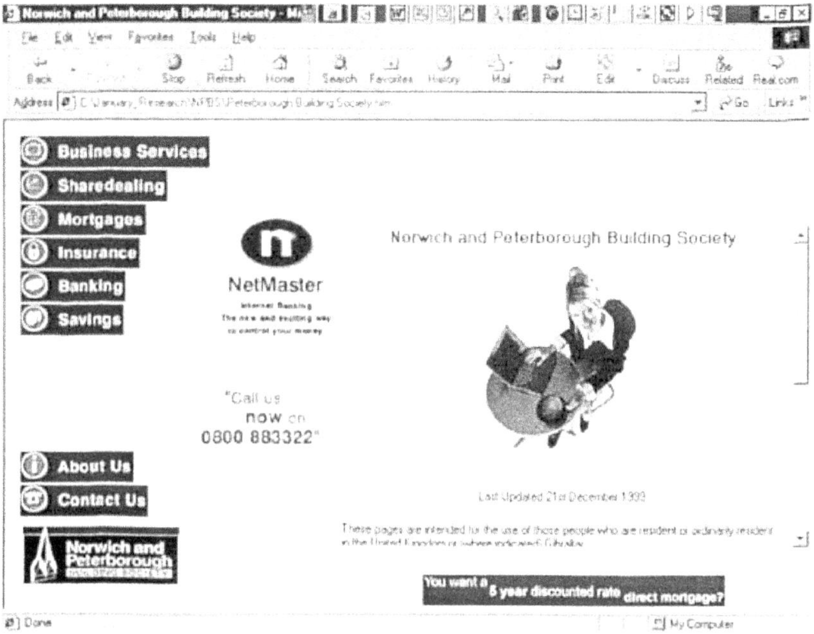

The Norwich and Peterborough Building Society was formed by the merger of the Peterborough and Norwich building societies m 1986 and operates 60 branches mainly concentrated in the eastern counties of England.

In terms appearance this Web site is simple. Some sections of the site appear to be quite dated - for instance. "There is no charge for Gold Current Account holders. Business Gold account holders will be charged £7.99 per month (collected directly from the account on or about the 17th day of each month). As a special introductory offer the NetMaster service will be offered FREE for the first six months". There is no mention when the six-month starts. The wording was identical when the site was viewed in mid 1999 and early 2000, and subsequently as late as June 2000.

During the navigation process, it does not open new browser widows depending on the subject heading. When submitting confidential information, such as in the case when a customer registers for online banking, these details are not submitted over a secure server. The section on guidance on choosing an account can be confusing due to the extremely a large selection of different accounts.

14.1 Demo and the Operation of the Account

Demonstration is quite simple and clear with seven separate screen shots. The NetMaster system was developed primarily for IBM system. However, if a user makes use of a MAC they may be able to access NetMaster.

Items on the account statement can be sorted according to, debit card payments, direct debits, standing orders, and cheques. Further these in turn can also be sorted according to ascending or descending order. Each customer is issued with a customer number and a 10 character NetMaster Access Password. Additional security is provided via an automatic lockout following a period of inactivity (5 minutes) and also after 3 failed attempts to access NetMaster.

14.2 Query/Email

The facility to communicate with the bank is neither prominently displayed nor made over a secure connection.

15. Royal Bank of Scotland (RBS)

URL - http://www.rbs.co.uk/pc&online/default.htm

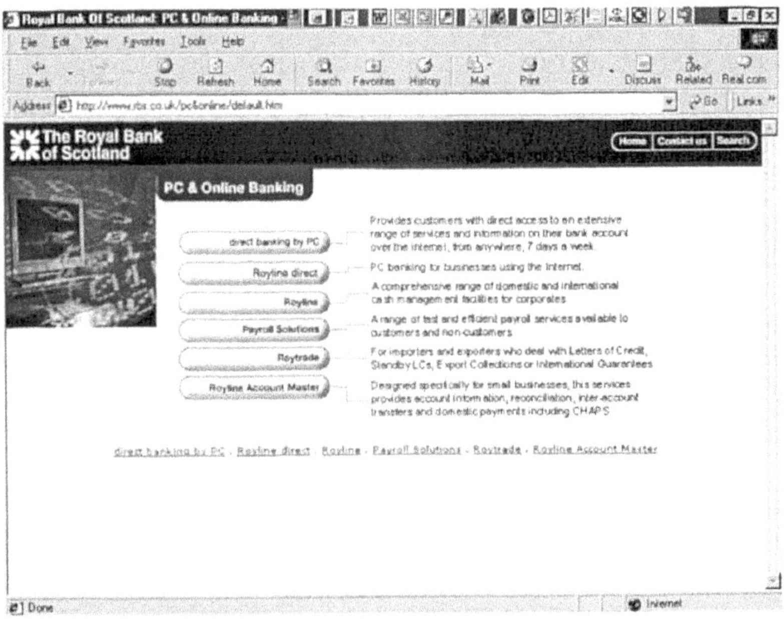

The RBS is one of the oldest banks in the UK, and has the distinction of having made the first overdraft in 1728. The bank operated 650 branches in the UK, prior to its acquisition of NatWest in mid 2000. RBS Claims to be the first direct banking by PC is the first to announce an Internet banking service in the UK.

However in light of the progress made by other banks, the RBS Web site do not appear to have made much incremental progress in terms of functionality, appearance, navigation, etc. It incorporates a free ISP service. In addition, the Web site contains some information of interest on inflation, economic trends, etc. FAQ section is quite extensive. Some sections of the site require the user's computer to accept cookies. RBS appears to have invested more heavily in promoting the Royline for Windows™. Which delivers cash management facilities, including euro, for corporate customers.

Site navigation can be difficult, one tends to get 'lost' within the site. Site map needs to be more coherent. Certain sections of the site are connected via a secure server. A new browser window opens only to display the site map. The search engine it not very effective. The Web site uses frames extensively.

The savings calculator performs calculations both backwards and forwards. For instance, to calculate the desired lump sum and the end date, select the savings account and then click the calculate button, it gives you the amount that needs to be saved on a monthly basis or the amount to be invested today.

15.1 Demo and the Operation of the Account

The demo is a collection of product screen shots. Appears to have invested relatively little in the development of the Internet banking services.

The system only works on Windows® 95 or 98 or Microsoft® Windows® NT 4.0. Direct banking by PC is not available to MAC, Windows® 3.1, UNIX or OS/2 users. RBS claims that the reasons are security ones, with these platforms not being able to accommodate equivalent security measures with ease.

To open a direct banking by one needs to have a Royal Bank of Scotland account and to be a customer of the telephone banking service, Direct Banking. The application form can be filled online, but has to be printed and mailed to RBS. Once the security details have been arranged the customer is able to use direct banking by PC. Up to 50 current and savings accounts can be managed online. The customer can assign names to accounts.

15.2 Query/Email

A Web based form is available for communication, but it is not connected over a secure server.

Appendix B

Date

Address

Dear Sir/Madam,

XXX: Internet Banking Service

We are presently carrying out a study examining purchase determinants for Internet banking services.

For this purpose we are collecting estimates of the number of Internet banking service users with each of the UK banks, from banks themselves and from journalists, etc. We understand that you are busy, therefore could you please consider filling in this box, and returning in the envelope provided. Any other information you could provide is most appreciated.

Launch of Internet Banking Service (date)	
Number of customers signed up as of 01.01.2000	
Number of active customers as of Ol.01.2000	
Average number of customers singing up with the service per week/month.	

Thanking you,
Yours faithfully,

Trevor Buck
Professor of Business Policy

INDEX

www.ingramcontent.com/pod-product-compliance
Ingram Content Group UK Ltd.
Pitfield, Milton Keynes, MK11 3LW, UK
UKHW021328190726
13845UKWH00004B/80

9 789349 964259